WITHDRAWN

Minerals, Rocks
and Inorganic Materials

*Monograph Series of Theoretical
and Experimental Studies*

2

Edited by

W. von Engelhardt, Tübingen · T. Hahn, Aachen

R. Roy, University Park, Pa.

J. W. Winchester, Tallahassee, Fla. · P. J. Wyllie, Chicago, Ill.

Edward Hansen

Strain Facies

*With 78 Figures
and 21 Plates*

Springer-Verlag New York · Heidelberg · Berlin 1971

64856

Dr. Edward Hansen
Yale University, New Haven, Conn.
and Geophysical Laboratory,
Carnegie Institution of Washington
Washington, D. C./U. S. A

Present Address:
State University College
New Paltz, New York/U. S. A.

ISBN 0-387-05204-6 Springer-Verlag New York Heidelberg Berlin
ISBN 3-540-05204-6 Springer-Verlag Berlin Heidelberg New York

This work is subject to copyright. All rights are reserved, whether the whole or part of the material is concerned, specifically those of translations, reprinting, re-use of illustrations, broadcasting, reproduction by photocopying machine or similar means, and storage in data banks.

Under § 54 of the German Copyright Law where copies are made for other than private use, a fee is payable to the publisher, the amount of the fee to be determined by agreement with the publisher.

The use of general descriptive names, trade names, trade marks, etc. in this publication, even if the former are not especially identified, is not to be taken as a sign that such names, as understood by the Trade Marks and Merchandise Marks Act, may accordingly be used freely by anyone.

© by Springer-Verlag Berlin · Heidelberg 1971. Library of Congress Catalog Card Number 72-89551. Printed in Germany. Typesetting, printing and bookbinding: Brühlsche Universitätsdruckerei Gießen

Foreword

This series of monographs represents continuation on an international basis of the previous series MINERALOGIE UND PETROGRAPHIE IN EINZELDARSTELLUNGEN, published by Springer-Verlag. The voluminous results arising from recent progress in pure and applied research increase the need for authoritative reviews but the standard scientific journals are unable to provide the space for them. By their very nature, text-books are unable to consider specific topics in depth and recent research methods and results often receive only cursory treatment. Advanced reference volumes are usually too detailed except for experts in the field. It is often very expensive to purchase a symposium volume or an "Advances in .." volume for the sake of a specific review chapter surrounded by unrelated chapters. We hope that this monograph series will by-pass these problems in fulfilling the need. The purpose of the series is to publish reviews and reports of carefully selected topics written by carefully selected authors, who are both good writers and experts in their scientific field. In general, the monographs will be concerned with the most recent research methods and results. The editors hope that the monographs will serve several functions, acting as supplements to existing text-books, guiding research workers, and providing the basis for advanced seminars.

March 1971

W. VON ENGELHARDT, Tübingen
T. HAHN, Aachen
R. ROY, University Park, Pa.
J. W. WINCHESTER, Tallahassee, Fla.
P. J. WYLLIE, Chicago, III.

Contents

I. Introduction

The idea that the present character of a rock expresses its conditions of formation is called the facies concept in geology and is used in different fields of the science for classification and interpretation. For example, it is generally accepted by stratigraphers and sedimentologists that the present aspect of a sedimentary rock (its particle size and shape, degree of sorting, bedding characteristics, etc.) is indicative of a particular depositional environment, and that sedimentary rocks displaying the same characteristics were deposited under similar environmental conditions. Another example is the mineral facies concept in metamorphic petrology. It has been shown from thermodynamic considerations that intensive parameters like temperature, total pressure, and oxygen fugacity determine the equilibrium mineral assemblage displayed by any rock of a given bulk chemical composition; conversely, rocks displaying the same mineral assemblage are interpreted to have attained equilibrium by recrystallization under the same ranges of these environmental parameters.

An attempt is made here to show that this is also an appropriate and powerful concept when applied to the strain features in rocks. In this application, structures are classified by their appearances in outcrop, and the classes, called *strain facies*, are interpreted to have formed within distinct strain environments; the same kind of strain environment produces structural features characteristic of a single strain facies. Thinking in this manner, one is able to see similarities in form and genesis of structures that would not generally be compared, for example the folds in a salt dome and the folds in a basin in the metamorphic core of a mountain chain, or the folds in a landslide of tundra sod and those associated with late-stage nappe emplacement in high-grade metamorphic zones. Furthermore, it permits interpretation of otherwise enigmatic structures, simply by virtue of their displaying the same form as others whose genesis may be known independently, perhaps because of better exposure or a more favorable structural situation.

The principal structures used here for classification by facies are mesoscopic[1] folds — mesoscopic because the facies concept was con-

---

[1] *Mesoscopic* is the scale of an outcrop or hand sample; it is intermediate between *microscopic,* the scale of a thin section, and *macroscopic,* the scale of a map compiled from outcrops *(Weiss* and *McIntyre,* 1957, p. 577; *Weiss,* 1959*b*, p. 7). Mesoscopic is not a synonym for "minor", as used by *Elliott* (1965, p. 866); rather, minor as opposed to "major" roughly comprises both the microscopic and mesoscopic scales.

ceived and is used for the properties of rocks in outcrop, and folds because they are the most sensitive of mesoscopic structures, at least in rocks deformed primarily by flow. The mesoscopic folds that form the basis of this study are found in the medium-grade metamorphic rocks in Trollheimen, Norway. Folds are well developed there, as many as three generations superimposed within certain outcrops, and interference domes and basins are not uncommon. Consequently, a good portion of this book is devoted to the description and analysis of Trollheimen's structures. They serve here as specific examples in the development and presentation of the general concept.

An adequate description of the features to be classified is of primary importance to the application of the facies concept. It is easy to describe the mineral assemblages of most metamorphic rocks, and stratigraphers are fairly well agreed upon the critical properties of sedimentary rocks, but unfortunately the art of describing folds is not so advanced. For example, most structural geologists still ignore the axial surfaces of folds in favor of their more obvious hinge lines, even though axial surfaces are the most fundamental manifestations of strain in certain types of folds and in certain structural situations. Nevertheless, the apparent individuality of practically every fold encountered in the field is probably the main reason why the description and classification of folds has not been a matter of great concern. A mineral assemblage must be contained in the single layer whose composition it represents, as must the properties of a sedimentary rock, obviously; but mesoscopic folds commonly span several layers of different rock types, and, because of the individual thicknesses and strengths of the layers, the folds vary in form from layer to layer and fold to fold. This problem can be ignored in Trollheimen, however, because one of the important rock units there, composed mainly of thick sequences of homogeneous, finely laminated quartz schist, is exposed in all the macroscopic structural complexes in the area. Therefore, it was possible to standardize the descriptions by restricting them to folds exposed in this one rock type, and only these are used to delineate the strain facies.

The folds are described by a set of twelve properties, of which eight are characteristics of an individual fold, and the remaining four are characteristics of a group of folds. Some of the properties that regularly appear in fold descriptions are included here, such as similar or parallel geometry, and straight or curved hinge lines; several properties, like the cylindricity of a fold, are redefined; and a few others, which incidentally turn out to be some of the more useful ones, are new to fold descriptions, such as the quantitative ratios of height to width and depth to width which express certain aspects of a fold's form. This particular set of twelve properties is used because it provides an adequate characterization of Trollheimen's fold types, and not because it is considered complete. These properties serve here as indices to facies affinities, and Trollheimen's mesoscopic folds, as well as their associated structures, are classified upon the basis of these properties into separate strain facies.

The critical step in showing that the facies concept is applicable to strain features is to demonstrate that structures classed by appearance into the same facies have indeed developed in a similar fashion and that those in different strain facies have in general developed differently. At present two kinds of independent evidence recorded in strain features are well enough understood that analysis of them can lead us to significant aspects of the development of the strain features: one is the orientation of slip lines; the other is the orientation of stress fields. Unfortunately, stress indicators such as twin lamellae in calcite and deformation lamellae in quartz are rare in the rocks of Trollheimen, and so the disposition of causative stress fields relative to the folds could not be deduced. Nevertheless, the geometry of multiple folding there is well suited to the study of slip-line orientations, and these, therefore, were used to link the strain facies to their modes of formation. A simple example of the relationship between structures and slip lines can be seen in a rug allowed to slide down an inclined plane under the force of gravity and flex into folds against some barrier like the floor. The fold axes would be horizontal and mutually parallel within a few degrees but approximately perpendicular to the down-dip movement line (or slip line), which in this experiment we know from observation.

Several methods of deducing slip-line orientations have been used during this study. The first of them is the "Weiss method," which is based on the geometry of passive lineations rotated during slip folding (*Weiss*, 1955, 1959a; *Ramsay*, 1960). Two additional methods, involving the plane of no folding (or hinge-line node) in superposed folds and the geometry of interference domes and basins, were developed in Trollheimen, though their basic theory can be found in the work of *Weiss* (1959a) and *Ramsay* (1962b), respectively. Furthermore, three new methods have emerged from consideration of the asymmetry of folds in more complex situations than have formerly been treated. Together, these methods are considerably more potent than any one of them alone; wherever possible, orientations of slip lines were deduced from several of them, thereby obtaining independent checks on the reliability of the solutions.

Strain environments of the facies are described in terms of simple, geometric flow types, inspired by *Mackin's* types of intrusive

1*

flow (*Mackin,* 1947). In general, the flow types are identified by a qualitative analysis of the strain relative to the slip-line orientations recorded by the structures. In our previous example, the rug allowed to slide down an inclined plane would have undergone a type of divergent flow as it crumpled against the floor. As another example, the structures in salt domes are generally agreed to have developed during a type of convergent flow and to have been modified later through rotation into a type of parallel, and in some places divergent, flow. The flow environments of the strain facies recognized in Trollheimen are similarly described, but with greater attention to detail.

The chapters that follow are grouped in three unequal parts. The first part consists of three chapters that are introductory or textlike in nature. They deal with the descriptive and interpretive systems mentioned in the foregoing paragraphs, namely the properties of folds, methods of deducing slip-line orientations, and types of flow. The second part, consisting of four chapters, is concerned with the rock units and structural complexes in southeastern Trollheimen. Three of these chapters are named for the three strain facies recognized in the area; in them are described the mesoscopic structures that make up each strain facies as well as the macroscopic structures that together constitute the structural complexes in which the facies occur. The third part, comprising the last chapter, is synthetic. Included in this chapter are a comparison of the strain facies of Trollheimen and a discussion of the general concept of strain facies.

Before proceeding to the systems used in this book, it seems appropriate to introduce the area for which they were designed.

Sketch of Trollheimen

Trollheimen is located approximately 100 kilometers southwest of Trondheim and 350 kilometers north and slightly west of Oslo (Fig. 1). Its mountains are gentle and covered with tundra and snowfields, and its valleys are U-shaped and broad, filled with birch, pine, and spruce. It is uninhabited for the most part, its valleys and lower slopes being used in the summer for dairy farming, and its peaks at Easter for cross-country skiing. Its more permanent inhabitants are lemmings, ermine, fox, and reindeer. The summer climate can be warm with short thundershowers in the afternoon when the wind comes from the south over Norway, but when the wind blows from the northwest off the North Atlantic, a blizzard can drop tens of centimeters of snow in three or four days. The geologists' field season is July and August, and only in a remarkable summer do possible working days average more than one out of every two.

The tectonic setting of Trollheimen is the metamorphic core of the Caledonides, which in this part of Norway trends N60°E. The great nappe complex of sparagmite and anorthosite-gabbro parallels the core zone and lies 100 kilometers southeast of Trollheimen. The

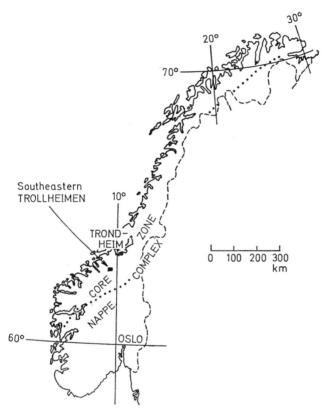

Fig. 1. Index map of Norway

core spans the distance from the nappe complex northwestward to the coast and is of the order of 180 kilometers wide in this region. Though little mapping has been done, reconnaissance work and local, more detailed studies show that it is not unlike the cores of other, better-known mountain belts with large domal and recumbent structures involving remobilized basement rocks as well as a metamorphosed cover sequence (Strand, 1960, 1961).

The rocks of Trollheimen have been metamorphosed to the almandine-amphibolite facies. Almandine, oligoclase, and microcline are abundant, but aluminous minerals more sensitive to this environmental range are scarce because the bulk compositions of the various rock types, such as granitic gneiss, quartzite, calcareous schist, and amphibolite, do not permit their appearance. The mineral pair kyanite-almandine has been found in three isolated localities. A rather large proportion of the rock column could yield a melt corresponding to the granite minimum (*Tuttle* and *Bowen*, 1958, p. 54), but no granitic intrusives are seen. Along strike to the west and southwest, beyond the area of this study, the metamorphic grade gradually increases, and granite, anorthosite, and eclogite appear. To the east and northeast the grade decreases sharply to the greenschists of the Trondheim basin, which forms Trollheimen's eastern border.

The combination of good exposure and easily mappable rock units fairly well eliminates configuration of the units as a problem. Continental ice sheets and valley glaciers have carved more than a kilometer of vertical relief and have left the bedrock in many areas devoid of overburden and practically unweathered. The major rock units are uniform within themselves, especially in the lower two-thirds of the column, and their contacts are sharp and clean, especially in the upper two-thirds of the column. The most important contact that traces the form of the macroscopic structures across the land surface is between gray-white feldspathic quartzite and dark green amphibolite; on clear days this line can be seen from mountain tops for as great a distance as 5 or 6 kilometers.

This study began with the geologic mapping of the 220 square kilometers shown in Plate 1 (in pocket). Parts of four field seasons were spent on this task, the amount of time totaling 8 months, of which at least 3 were lost because of adverse weather conditions. Consequently, the area covered by Plate 1 has been mapped with varying degrees of attention. The northeastern third has been done in reconnaissance on topographic base maps at a scale of 1 to 100,000, and the remaining two-thirds by mapping with average attention to detail on a topographic base of 1 to 25,000. A small portion of the western area has been mapped with a plane table and alidade at scales of 1 to 100 and 1 to 20. The study of mesoscopic structures also shared this variation in focus. The area immediately surrounding Kam tjern in the southwest, around the cabin used as operational headquarters for three of the summers, and the area of Svahø and Lang tjern in the southeast, near the ski lodge used during the fourth summer, have been studied in greatest detail.

Well over 3000 measurements of linear and planar structures were made in Trollheimen during the course of this study. Nevertheless, only the measurements considered representative of regional trends or critical to the macroscopic structure are recorded on the geologic map; they number about 250. Addition of the remaining readings to the map would only serve to make it more crowded and less readable with little additional illumination of the macroscopic structure. One might argue that by so limiting the structural data on the map the reader is unable to judge from the density of symbols where the exposure is good and where the mesoscopic structures are best developed. However, the lines locating contacts between rock units indicate the quality of the exposure, and a special map included in the text has been constructed to illustrate the distribution and intensity of development of the mesoscopic structures. Moreover, most of the several thousand measurements are pertinent to the discussions of the mesoscopic structures themselves and consequently are presented in histograms, maps, and spherical projections within the body of the book.

Acknowledgments

I am grateful to *John Rodgers* for inspiration, continued encouragement, and stimulating discussion as teacher, critic, and friend throughout all stages of this study. Field work was generously supported by Norges Geologiske Undersøkelse under the direction of *Harald Bjørlykke*, by the Department of Geology at Yale University, and by the Carnegie Institution of Washington. The Kgl. Norske Videnskabers Selskab Museum under the direction of *Erling Sivertsen* kindly permitted use of Kamtjernhytta as base for three summers' field work. The work was carried out with the able assistance of *S. W. Carmalt, John de Neufville, John Lucas, G. H. Myer,* and *R. J. Twiss.* Important discussion was provided by *B. C. Burchfiel, S. W. Carey, J. C. Maxwell, W. H. Scott, R. S. Stanley,* and *R. J. Twiss. F. A. Donath, F. A. Hills, R. B. Parker, L. B. Platt, John Rodgers, W. H. Scott, R. S. Stanley, M. S. Walton, Horace Winchell,* and *Rosemary Vidale* critically reviewed the manuscript; *Larry Finger, G. W. Fisher, Keith Howard, S. W. Richardson,* and *R. J. Twiss* reviewed portions thereof. To all these people, I extend my sincere thanks.

II. Twelve Properties of Mesoscopic Folds

For the present purpose it is necessary to be able to describe some of the essential features in the total aspect, or *style*, of any group of folds. The properties used in this work to characterize the styles of Trollheimen's folds are discussed here in order to avoid introducing them at various points within the presentation of the strain facies. This is, in effect, a chapter on the fold terminology used throughout this book.

Definitions

A *fold* is the distortion of a volume of material that manifests itself as a bend or nest of bends in linear and planar elements within the material. A bend in one surface is only part of a fold if it is nested in bends of adjacent surfaces; the whole fold involves all the surfaces within the volume so distorted. The *fold axis* is the line which, when moved parallel to itself in space, generates the form of the fold (after *Wegmann*, 1929, p. 102, and *Clark* and *McIntyre*, 1951, p. 594). A *hinge line* is the line along which maximum curvature of a folded surface occurs (*Turner* and *Weiss*, 1963, p. 106); the *limbs* are the remaining portions of the folded surface on both sides of the hinge line. The *axial surface* is the surface that contains all the hinge lines of a fold (*Donath* and *Parker*, 1964, p. 48). A *profile* of a fold is a section cut perpendicular to the fold axis (after *Wegmann*, 1929, p. 107, and *McIntyre*, 1950, p. 331).

Only two principal mechanisms of folding need be considered here: *Slip folding* is the folding produced by slip between submicroscopic and mesoscopic layers inclined to passive compositional layers (*Knopf* and *Ingerson*, 1938, pp. 157—159; *Weiss*, 1959a, p. 92; *Donath* and *Parker*, 1964, p. 48). "Flow folding" is included in this mechanism, whether or not the slip or flow occurs along discrete surfaces (*Carey*, 1954, pp. 92—98). *Flexural-slip folding* is the flexing of individual, active, compositional layers and the simultaneous slip between those layers that together produce a fold *(Knopf* and *Ingerson*, 1938, pp. 159—160; *Weiss*, 1959a, p. 92; *Donath* and *Parker*, 1964, p. 49).

Properties of an Individual Fold

Let us consider a fold in profile:

1. Type of Geometry

If the lines or bands that define the fold mimic each other throughout the fold, the type of geometry is *similar (Billings, 1954, p. 56, Fig. 42A)*; in this type, the lines are closer together and the bands are narrower on the limbs than at the hinge. If the lines and bands parallel each other throughout the fold, the type is *parallel (Billings, 1954, p. 56, Fig. 42B)*; no apparent narrowing on the limbs or widening at the hinge is seen. A single fold may in different parts exhibit properties of both types, but if it approaches one more than the other, it is designated as that type.

2. Nature of Hinge and Limbs

The lines and bands of a fold may describe a *broad* curve at the hinge, or the hinge may be *sharp* and lack any appreciable curvature (cf. *Billings, 1954, p. 42, Fig. 28A*). Similarly, the limbs may be *broadly curved,* or they may be *straight.*

3. Ratio of Short-Limb Height to Width

The terms amplitude and wavelength are defined for the geometry of regular wave trains, such as the sinusoidal transverse wave drawn in Fig. 2*a.* The amplitude, *A,* is half the distance measured

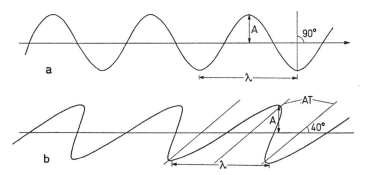

Fig. 2. Amplitude *(A)* and wavelength (λ) of a sinusoidal transverse wave *(a)* and asymmetric folds in profile *(b)*; traces of axial surfaces labelled *AT*

from a crest to a trough at right angles to the direction of propagation. It parallels the line of symmetry that bisects a crest or a trough into two equal halves, and it measures the maximum amount of particle displacement within the path of displacement. The wavelength, λ, is the distance measured between adjacent crests or adjacent troughs parallel to the propagation direction. It is a measure of the distribution of the disturbance along the line of propagation.

In many areas of deformed sedimentary rocks, the shapes of folds make the analogy to regular wave trains obvious, and they are easily described by measurements of amplitude and wavelength. Folds in metamorphic rocks, however, are not so easily described. Traces of their axial surfaces, *AT*, commonly make acute angles with the mean orientation of the folded line[2] (Fig. 2b). The amplitude, therefore, does not measure the maximum displacement in the path of a displaced particle of the fold. Instead, it combines the maximum displacement of a particle and the angular orientation of the path taken by a displaced particle into a single measurement. In some folds the path of displacement can be assumed to parallel the axial trace, but in many, it has some other orientation which may not be known. Therefore, an amplitude measurement on folds such as these gives information combined in a way that can be difficult to evaluate.

The wavelength is a measure of the distribution of folds in a line. If folds are few and spaced far apart, the wavelength is large; conversely, if many folds occur close together, the wavelength is small. However, if folding is irregular, and each fold has its own size and shape and distance from the others — as is commonly the situation — a fold has one wavelength measured to the crest of the next fold on the left and a different wavelength measured to the fold on the right, so that the analogy to regular wave trains is strained.

To solve these problems, *Ramsay* has recently redefined these measurements as applied to folds. Amplitude is the distance between an inflection point in a folded line and one of the two envelopes to that line, and wavelength is the distance between alternating inflection points in the folded line; these distances are measured at right angles to each other but parallel to any specified pair of reference axes within a profile section (*Ramsay*, 1967, pp. 351—354, Fig. 7 - 8). An alternative solution, one that is used here, is to abandon amplitude and wavelength and take measurements of a slightly different nature:

Where the trace of the axial surface is not perpendicular to the mean orientation of the line that describes the folds, an asymmetry of the folds and a consequent coupling of hinges result, two hinges being close together and separated by a greater distance from the next two hinges in either direction along the same line (Fig. 2b). (Folds at the hinge of a larger fold, where the axial trace is nearly perpendicular to the mean orientation of the line, are an exception to

---

[2] *Ramsay* (1967, pp. 351—353, Fig. 7—9) uses the term *sheet-dip* for the mean orientation of a folded surface within a given area.

this generalization; they are symmetrical, their hinges may be evenly spaced, and they approach the form of the wave pictured in Fig. 2a.) One of the coupled folds shown in Fig. 2b has been magnified and redrawn in Fig. 3a. A *short limb* joins the two hinges, and the *long limbs* extend away to the hinges of the next pair (*White* and *Jahns*, 1950, p. 197). The distance between the two hinges coupled by the short limb, measured parallel to the axial trace, is the *short-limb height (Matthews*, 1958, pp. 512–513), labeled *H* in the figure. The

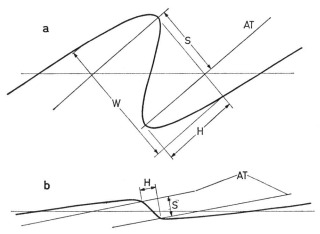

Fig. 3. Height *(H)*, width *(W)*, and spacing *(S)* of coupled folds in profile; traces of axial surfaces labelled *AT*

short-limb height measures the length along the axial trace that a line is folded back on itself, though it may or may not parallel the path of particles displaced by the folding. Two lines can be constructed perpendicular to the axial trace, each tangent to the fold at a hinge and intersecting the more distant long limb. The *width* (W) is the projected distance between the points of intersection of these lines with the long limbs, measured perpendicular to the axial trace. It is a measure of the distance across the axial trace within which the line is folded.

A fold in outcrop rarely involves only a single line (the trace of one bedding surface, for example); ordinarily it involves many. These two distances are measured, however, on a single line chosen from the many lines so that the value of the short-limb height is as large as possible. (Commonly the width attains its maximum value on the same line.) A set of measurements so taken has been found to represent adequately the height and width of a fold in outcrop.

The height and width are as fundamental measurements of the dimensions of a fold as are the amplitude and wavelength as redefined by *Ramsay*. They are somewhat more practical for the field geologist because they do not require locating the two inflection points on the long limbs of the fold or locating one of the envelopes and because less of the folded surface need be exposed to take them. Amplitude and wavelength could not be measured with confidence on either of the lines shown in Fig. 3, for example. Of course, depending upon the purpose of the study, either or both sets of measurements may be useful.

The absolute size of a fold is not an element of style — at least not as envisaged here. Rather, folds of different sizes may exhibit the same structural elements and thus have an identical style. In order to eliminate absolute size and express only the form of a fold, ratios of absolute measures are needed. The ratio of short-limb height to width (H/W) is useful in this respect, because it expresses quantitatively the relative amount of overfolding, or doubling of a line back on itself, regardless of size.

In a series of coupled folds with different height-width ratios, the width approaches the spacing (S) between the axial traces *(Matthews*, 1958, pp. 512–513) as the short-limb height approaches zero. At zero height, the width does not exist. In a fold in which there is no doubling of the folded line back on itself parallel with the axial trace, as in Fig. 3*b*, the height has a negative value, and the spacing S replaces the width; the height-spacing ratio is negative. This situation occurs in similar folds in which the axial traces make small angles with the mean orientations of the folded lines. Approximately one such fold in fifty was encountered in Trollheimen. Nevertheless, in any study that attempts to treat the dimensions and forms of folds quantitatively, folds such as these cannot be ignored.

Early in the field work associated with this study, it was noticed that the height-width ratios of similar folds in which H/W was less than 0.3 were difficult to determine accurately on the outcrop. An arbitrary decision was made at that time to assign the height-width ratios of all such open folds, including those with negative heights, the positive value of 0.1. This decision has been followed to the present, and the data of H/W given for similar folds in subsequent chapters include such arbitrary values to the amount of about 3 percent. An alternative would have been to measure height-spacing ratios, the range of which, from positive through negative values, includes all folds.

If the ratio of height to width is greater than 1.0, the short limb nearly parallels the long limbs, and the fold is nearly isoclinal. If the ratio is less than 1.0, the attitude of the short limb is different enough from that of the long limbs so that the fold is open. A series of folds with height-width ratios ranging from 0.1 to 5.0 is shown in Fig. 4.

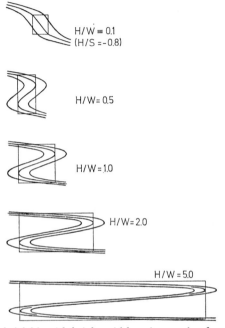

Fig. 4. Coupled folds with height-width ratios ranging from 0.1 to 5.0

4. Ratio of Depth to Width

The trace of the axial surface of a fold in profile continues as far as the fold continues. Therefore, the length of the axial trace, or *depth* (*D*), measures the disturbance by folding parallel to the axial trace, just as the width measures the disturbance perpendicular to the axial trace (Fig. 5). The ratio of the depth to the width of a fold (D/W, where W is measured at maximum H) expresses the elongation of the fold in profile. One can also think of this ratio as a measure of the harmony of a fold. A fold with a large ratio of depth to width is harmonic; the change in form from a line involved in folding to an undisturbed line is subtle. If a fold has a small ratio, however, changes in form from line to line are striking, and the fold is relatively disharmonic. Fig. 6 illustrates folds with ratios of depth to width ranging from 1 to 15.

Let us now consider a fold in three dimensions by adding the dimension of the hinge line:

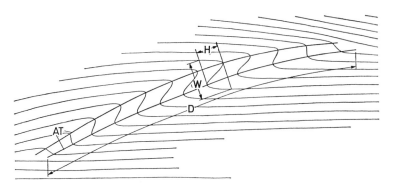

Fig. 5. Depth *(D)* of a coupled fold in profile; height *(H)*, width *(W)*, and traces of axial surfaces *(AT)*

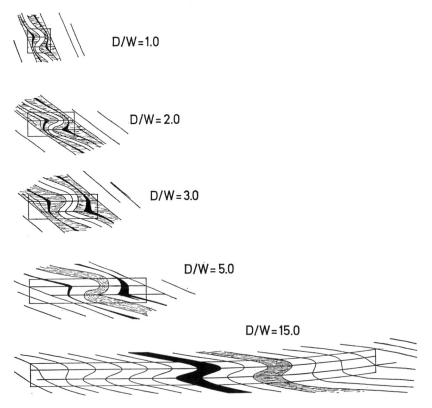

Fig. 6. Coupled folds with depth-width ratios ranging from 1.0 to 15.0

5. Length and Character of Hinge Line

Taking the axis of a fold as fabric *b*, the axial surface as *ab* and its pole as *c* *(Sander,* 1926, p. 328; *Weiss,* 1955, p. 229), the five basic dimensional parameters of a coupled fold are the short-limb height and the depth, which are measured parallel with fabric *a*, the length of the hinge line, measured parallel with *b*, and the width and spacing, measured parallel with *c*. The form of the fold can be expressed by the ratios of these parameters.

Unfortunately, if a fold is exposed, part of it has been removed, and therefore all five of these distances cannot be measured or even approximated. Exposure surfaces cut most folds at some angle to the fold axes, not parallel to them, with the result that the most difficult parameter to measure is the length of the hinge line. Consequently, during the present work, the length of the hinge line has been studied qualitatively. If a hinge line is seen to begin or end in the outcrop, it is considered *short*, and if there is no indication of a beginning or an end, it is *long*. In the same manner, the character of the hinge line is described as *straight* or *curved*.

6. Cylindricity

Cylindricity is used here to designate the degree to which the form of a folded surface at any given profile approaches a cylinder. A cylinder in geometry is a surface described by any straight line moving parallel to itself through space. Cylindricity, therefore, is a measure of how well the form of a fold can be described by the fold axis moving parallel to itself through space.

A *cylindrical* fold at any given profile has the form of a cylinder and can be described completely by the fold axis. In other words, a cylindrical fold can be defined as a fold from which all attitudes of surfaces, measured at a given profile and projected on a sphere, intersect at a point; if poles to the surfaces are plotted, they lie on a great circle of the sphere (Fig. 7*a*). A *cylindroidal* fold has more or less the form of a cylinder at any given profile. In reference to a spherical projection, it is a fold whose surface attitudes at a given profile do not intersect at a point, but have a very small polygon of intersections, and whose poles to the surfaces lie within 5° of a single great circle (Fig. 7*b*). An *irregular* fold is far from cylindrical. Its surface attitudes, measured at a given profile, have a large polygon of intersections in spherical projection, and poles to the surfaces lie more than 5° from a single great circle (Fig. 7*c*).

The final category of cylindricity that is useful in describing Trollheimen's folds is *conical*. The profile of a conical fold changes in size, but not shape, along its hinge line. A conical fold has no fold axis, at least as used here, but its form can be generated by a line passing through a fixed point (*Turner* and *Weiss*, 1963, p. 108); nevertheless, the attitude of the cone axis, in the absence of a fold axis, can be used to describe its attitude. The form of a conical fold approaches

a cylindrical

b cylindroidal

c irregular

d conical

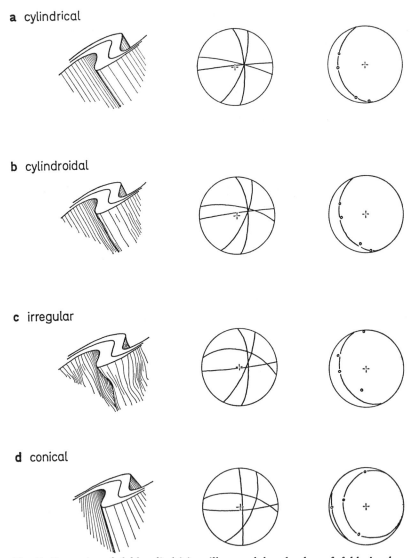

Fig. 7. Categories of fold cylindricity, illustrated by sketches of folds in three dimensions and by equal-area projections of attitudes of the folded surfaces, plotted both as great circles and as poles

a cylinder as its semiapical angle, measured between the cone axis and the line that generates the form of the fold, approaches zero *(Dahlstrom, 1954, p. 145)*. In spherical projection, the surface attitudes of a conical fold at a given profile describe a large polygon of intersections, and if poles to the surfaces are plotted they approximate a small circle (Fig. 7*d*).

7. Relation to Cleavage

"*Cleavage*, sometimes called *rock cleavage* in order to distinguish it from *mineral cleavage,* is the property of rocks whereby they break along parallel surfaces of secondary origin" *(Billings, 1954, p. 336)*. *Schistosity* is the type of rock cleavage determined by the preferred planar orientation of cleavage surfaces of crystals. *Slaty cleavage* can be considered the equivalent of schistosity in very fine-grained rocks. *Slip cleavage* is rock cleavage that shows offsets in the compositional layering across a cleavage surface. *Fracture cleavage* is rock cleavage that is relatively widely spaced (visible on the mesoscopic scale) but is not determined by the preferred planar orientation of dimensional axes of crystals. *Foliation* is used here to describe schistosity and compositional layering only where they are mutually parallel.

Cleavage parallels the axial surfaces of some folds. In other folds, one cleavage is folded, and a second cleavage is developed parallel to their axial surfaces. The relation of a fold to the cleavage(s) has many variations and is here considered a useful descriptive characteristic.

8. Relation to Mineral Lineation

Folds also exhibit different relations to the lineation defined by the preferred parallel orientation of long axes of minerals and by trains of minerals. For example, in one fold the fold axis may rigidly parallel the mineral lineation, in another the lineation may be folded, and, in a third, one mineral lineation may be folded and the fold axis may parallel a second lineation.

Properties of a Group of Folds

Folds of only a single generation are considered in this section.

9. Mean and Standard Deviation of Ratios of Short-Limb Height to Width

Seven hypothetical folds are shown in profile in Fig. 8*a*. Portions are drawn in solid lines where they are exposed and dotted lines

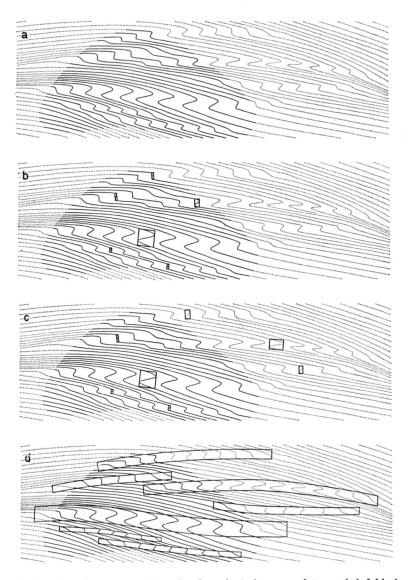

Fig. 8. Schematic representation of a hypothetical group of 7 coupled folds in profile; lines displaying the folds are solid where exposed and dotted where covered. *a* Folds traced by compositional layering. *b* Maximum height-width ratios of exposed portions of folds. *c* Maximum height-width ratios of both covered and exposed portions of folds. *d* Depth-width ratios of folds

where covered. Measurements of height and width can be made on the exposed parts of each fold, and the mean and standard deviation of their ratios can be computed for the group. The *arithmetic mean* or average ($\bar{r}$) is the sum of the ratios of height to width divided by the total number of ratios used. In the outcrop illustrated (Fig. 8*b*), height-width ratios measured where the heights are largest are (from top to bottom) 0.3, 0.3, 0.6, 0.1, 1.0, 0.5, and 0.5, and the mean is 0.5. If it were possible to measure the height-width ratio for each fold where the short-limb height is at a maximum by measuring the un-exposed portions where necessary (Fig. 8*c*), the ratios would be 0.5, 0.3, 1.3, 0.5, 1.0, 0.5, and 0.5, and the *maximum mean* (i. e. the mean of ratios obtained at maximum *H*) is 0.7.

Three factors — the absolute size of the folds, their height-width ratios, and their depth-width ratios — contribute to the difference between the maximum mean and the mean of ratios measured within any given exposure. (1) In general, the larger the folds, the smaller the percentage of their exposure, and the smaller the chance of sampling their height-width ratios at maximum height. (2) In folds whose heights are more than twice their widths ($H/W > 2$), the measured mean tends to differ from the maximum mean by a large amount because the spread in ratios along the axial traces is large and the chances of seeing the parts with low ratios are better than of seeing those with high ones. (3) In folds whose depths are several times larger than their widths ($D/W > 4$, Fig. 8), it is seldom possible to collect the values at maximum height from the whole group of folds. If exposure is good, the measurable mean may be close to the maximum mean, but if exposure is poor, the mean of measured ratios may be quite different from the maximum mean. Nevertheless, the mean of height-width ratios obtained from an outcrop is a minimum value for the amount of overfolding of the whole group of folds.

The standard deviation, *s*, for *n* ratios, $r_1, r_2, \ldots, r_n$, is given by the expression

$$ s = \sqrt{\sum_{i=1}^{n} \frac{(r_i - \bar{r})^2}{(n-1)}} $$

where $\bar{r}$ is the mean of the ratios (cf. *Davies*, 1957, p. 13). The standard deviation calculated for the exposed height-width ratios shown in Fig. 8*b* is 0.29, and for the maximum ratios, unexposed and exposed (Fig. 8*c*), it is 0.36.

As the example has shown, cover can alter the obtainable standard deviation from the "true" value which might otherwise be calculated from the maximum height-width ratios. However, even where exposure is not a variable, the standard deviation obtained

2*

from a group of folds can differ from the "true" standard deviation as the mean of the ratios varies. In two areas of equal exposure, the standard deviation of the group of folds with the lower mean of height-width ratios will tend to approximate the "true" standard deviation more closely than will the standard deviation of the folds with the higher mean, simply because there is a smaller range of possible variations in the ratios of exposed portions of folds when the mean is low. Depending upon the particular group of folds and the particular outcrop under consideration, the standard deviation obtained may be larger or smaller than the "true" standard deviation.

The standard deviation of height-width ratios gives the spread of values of overfolding in a group of folds. Ideally, where the exposure is good, the standard deviation measures the uniformity of folding. If each individual of a group of folds in nearly the same as the others with regard to the amount of overfolding exhibited, the standard deviation calculated for the group is small. If there is great variation in the amount of overfolding displayed by the folds, and each individual is quite different from any other, the standard deviation is large.

10. Mean of Ratios of Depth to Width

The two lowest folds in Fig. 8 are completely exposed, and their depths can be measured. Their ratios of depth to width are 20 and 24 (Fig. 8d), and the mean of these two ratios is 22. The other five folds are partly covered, but, if they could be measured and their depth-width ratios determined, we would find them to be 18, 13, 23, 18, and 16. The mean calculated for the group of seven folds is 19, which differs by 3 from the mean obtained for the two exposed folds.

It is probable that in a group of folds the shorter depths will be exposed and the longer ones covered. Nevertheless, this need not bias the sample; if the widths are proportional to the depths, the mean will remain essentially unchanged. The folds in Fig. 8 illustrate this point; where the depth is large the width is large, and where it is small the width is proportionally small. Where a group of folds has a small mean of depth-width ratios (about 3), more folds are likely to be completely exposed than where the group has a large mean of depth-width ratios (about 20). This effect does not change the values obtained; instead it changes the reliability of the values. It causes the small mean to be based on more observations than the large mean and to have a better chance of representing the "true" mean.

The mean of ratios of depth to width is a measure of the elongation of disturbance by folding in the planes perpendicular to the fold axes in the whole group of folds.

11. Preferred Orientation of Fold Axes

In addition to the usual map presentation, it is customary in structural studies to eliminate location and present the orientation of fold axes by projection from the lower hemisphere to equal-area nets, oriented within the horizontal plane and labeled with geographical coordinates. Any preferred orientations thus become apparent as point-maximum or girdle distributions, which may or may not be emphasized by contouring. This convention is followed here.

12. Asymmetry

There are two possible patterns made by the alternation of long and short limbs, when viewed in a section that cuts the hinge lines, and one is the mirror image of the other. For ease of equal-area net

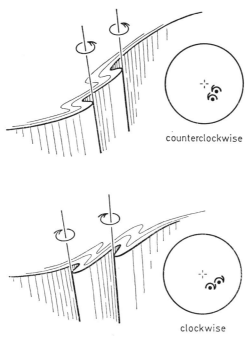

Fig. 9. The asymmetry patterns of folds, and their notation in spherical projection

presentation, these fold patterns are described here in terms of rotation of short limbs relative to long limbs about the hinge lines. As one looks at a fold *down* the plunge of its hinge line, its asymmetry is *clockwise* if the short limb is rotated in a clockwise direction relative to the long limbs; its asymmetry is *counterclockwise* if the short limb is rotated in a counterclockwise direction relative to the long limbs (Fig. 9). Clockwise folds correspond to the dextral pattern of *White* and *Jahns* (1950, p. 197), and counterclockwise to the sinistral. The usage is purely descriptive.

The asymmetry of a group of folds whose axial surfaces are spaced fairly equally is governed by the angle at which the axial surfaces intersect the mean orientation of the compositional layering,

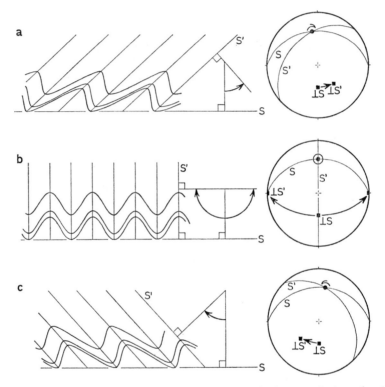

Fig. 10. Cross sections and equal-area projections in the horizontal plane showing the relationships between the mean orientation of the compositional layering, *S*, and the axial planes, *S'*, that govern the asymmetry patterns of folds. *a* Asymmetrical folds, clockwise. *b* Symmetrical folds. *c* Asymmetrical folds, counterclockwise

with reference to the horizontal plane by convention (Fig. 10). If one draws on a lower-hemisphere equal-area projection an arrow along a great circle in the short direction (<90°) from the pole to the mean orientation of compositional layering (⊥ *S*) toward the pole to the axial surfaces ($\perp S'$), the arrow points in a counterclockwise direction if the folds are clockwise, and it points clockwise if the folds are counterclockwise (Fig. 10*a, c*). If the layering and axial surfaces are mutually perpendicular, the folds are symmetrical, and the arrow measures 90° in either direction (Fig. 10*b*). In this situation, slight irregularities in the spacing of the axial surfaces may cause some of the folds to deviate from being symmetrical; however, this effect is normally random, the resultant folds of clockwise asymmetry are about as numerous as those of counterclockwise asymmetry, and the group of folds can be considered statistically symmetrical. In groups of folds whose axial surfaces are markedly unequally spaced, as in certain types of flexural-slip folds, the axial surfaces are coupled in such a way that the asymmetry relationships shown in Fig. 10 still obtain. Exceptions to this generalization are rare if they do exist.

Throughout this study, rules of order and location have been followed when describing the asymmetry of a group of folds: In an

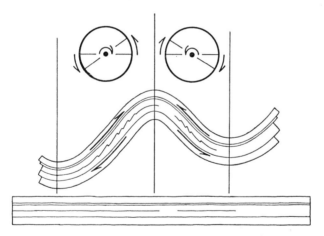

Fig. 11. Schematic representation of an interlayered sequence (bottom) shortened by flexural-slip folding (center). The high-order folds (center) developed in response to secondary shear couples generated by slip between layers on the limbs of the low-order fold. Spherical projections (top) show the overall sense of rotation of layering on each limb of the low-order fold (external arrows) and the asymmetry patterns of the secondary, high-order folds (internal, semicircular arrows)

outcrop where a low-order fold displays medium-order folds on its limbs and the medium-order folds have high-order folds on their limbs (*van Hise* and *Leith*, 1911, p. 123; *Stanley*, 1964, pp. 63—64, Fig. 8; *Ramsay*, 1967, pp. 354—355, Fig. 7-11), the asymmetry patterns of folds of only a single order are studied together. Furthermore, the asymmetry patterns of folds contained only in the volume of material confined between two adjacent axial surfaces of the next lower-order folds are studied together.

Most groups of folds whether produced by slip or flexural slip but limited according to the rules of order and location, display the same asymmetry relationship to lower-order folds. The asymmetry of a group of folds is *normal* if, as one stands on the axial surface of the next lower-order fold and faces toward closure, the folds are located to the left and are clockwise or to the right and are counterclockwise (Figs. 11, 12*a*). This normal asymmetry of mesoscopic folds

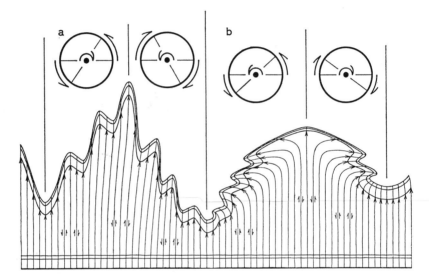

Fig. 12. Schematic representation of layers (bottom) folded passively in differential flow (center). High-order folds on the limbs of the low-order folds (center) are primary, having resulted from the same flow that produced the low-order folds. Spherical projections (top) show the overall sense of rotation of layering on each limb of the low-order folds (external arrows) and the asymmetry patterns of the high-order folds (internal, semicircular arrows). *a* High-order folds showing normal asymmetry relationship to low-order fold. *b* High-order folds showing reverse asymmetry relationship to low-order fold

is commonly used in field mapping to predict which way the trace of a rock unit has been offset by macroscopic folding *(Billings,* 1954, pp. 78–82); the use is restricted generally to folds with linear preferred orientations of axes. It is noteworthy that the normal asymmetry of a group of folds is opposite to the sense of rotation displayed by the limb of the lower-order fold on which the group is located (Figs. 11, 12a). The asymmetry of a group of folds is *reverse* if the relationship to lower-order folds is opposite to the normal relationship (Fig. 12b; *Skehan*, 1961, pp. 104—105). Finally, the asymmetry is *mixed* if both the normal and reverse relationships occur together.

Names of Folds According to Style

The strain-facies concept is founded on the observation that discrete differences in fold style exist within a given rock type. Trollheimen's three types of mesoscopic folds, which serve to demonstrate the concept, are separated and recognized wholly on the basis of style. These types of folds are common in folded belts around the world and, to date, have received extensive study by many workers. Practically every structural geologist would recognize them and have his own set of informal names for them. Such names are often based upon the local paragenetic sequence of the folds within a given area $(f_1, f_2, \ldots, f_n)$, upon their geometric relationships to local macroscopic structures (e. g., nappe or dome folds), upon certain prominent descriptive properties of the folds (similar, chevron, disharmonic), or upon some aspect of their inferred geneses ("a" folds, drag folds, flexural-slip folds). Despite such a variety of names, none of them designates a group of folds by its total aspect or style, which is the essential designation in the present context.

Portions of three subsequent chapters are spent describing and comparing Trollheimen's fold types. To facilitate discussion of these fold types and to formalize their separation on the basis of style, they are given names that denote their styles only. The point to be emphasized by this procedure is the style designation; the particular names are unimportant.

The style of a group of folds, as meant here, encompasses all the descriptive properties of the individuals within the group, as well as the properties of the group as a whole. In this book, the list of twelve properties discussed in the foregoing pages is used to delineate the styles. Thus a name given to the style of a group of folds becomes, in effect, a second-order descriptive term that denotes a combination of first-order terms, like similar, isoclinal, harmonic, and cylindrical. Furthermore, although the styles are described only as they appear in a single rock type (with certain minor exceptions), each style name includes the variations in style encountered in different rock types. Thus a style name denotes a particular combination of ranges of descriptive properties within a single rock type (e.g., H/W from 0.1 to 2.0, D/W from 6 to 15, cylindricity from cylindroidal to irregular, etc.) and a combination of somewhat different ranges of those properties within another rock type; and it denotes a combination of ranges expanded to include all rock types and interlayering thereof. In other words, a fold type described in one kind of rock goes by the same style name in another kind of rock, even though it may look slightly different.

With the idea in mind that if new names must be added to the literature they should at least be well chosen, an amount of effort disproportionately large relative to the significance of the names has been spent in naming these fold types during the past seven years. Letter designations and locality names were tried; names were formed by joining syllables of words describing important properties of the folds; and words were coined from nonsense syllables. The style names used in this book are merely the last and current set in a long series of sets that have fallen under the criticism of numerous colleagues. No doubt, given enough time, the current set would similarly disappear. Nevertheless, it now appears that a set of totally acceptable names for fold styles is not likely to be found, and that the current set is probably as good (and as bad) as any to be found in the future.

The style names used here are "sahlfold," "norfold," and "discfold." The first and third were formed by fusing the first letter of four of the more diagnostic descriptive properties of the folds and adding them to "fold". Exactly what properties supplied the letters are not identified here because all the properties are intended to be designated by the names, and not just those that yielded the syllables. The second name is a combination of "fold" and a syllable that is euphonious but otherwise meaningless with "fold."

III. Methods of Deducing Slip-Line Orientations from the Geometry of Folds[3]

One of the fundamental elements of the movement history of a deformed rock is the path of relative displacement of adjacent constituent particles during deformation. The orientation of such *slip lines* relative to the fabric elements of various groups of mesoscopic structures and to macroscopic structural complexes is of principal concern to this study. The methods of gathering this information that have been used in Trollheimen are described here in detail.

It is useful at this point to identify the slip of interest here with respect to flexural-slip folding and to contrast it with that in slip folding. The slip in slip folding can be described as penetrative; it occurs throughout the body of material involved in folding. The slip in flexural-slip folding is nonpenetrative, involving only the particles along descrete surfaces of layers or beds within the material; particles between the surfaces (within the layers) move in the totally different process of flexing. However, a second kind of nonpenetrative slip accompanies flexural-slip folding. It can be described as the net relative displacement between the layers involved in flexing and the adjacent layers that do not flex. It may or may not parallel the first kind of slip that occurs between the flexing layers. In this book, the slip line of interest with regard to a group of flexural or flexural-slip folds is of the second kind — the direction of displacement between the flexing layers as a mass and their surroundings.

Slip Folding of a Preexisting Straight Lineation

During folding by the flexural-slip mechanism, the rotation path of a preexisting straight lineation describes a cone in space (small circle in spherical projection), the axis of which parallels the fold axis. From fold to fold, the semiapical angle may vary from 0°, where the lineation parallels the fold axis and is therefore not rotated, to 90°, where the lineation is perpendicular to the fold axis and its rotation path is planar *(Weiss, 1959a, pp. 98–100; Ramsay, 1960, p. 76)*. During folding by the slip mechanism, a preexisting straight lineation is rotated in a planar path (great circle in spherical projection) toward parallelism with the slip lines. Therefore, the line defined by

---

[3] Most of this chapter appeared in *Carnegie Institution of Washington Year Book 65 (Hansen, 1967a)*.

the intersection of a slip plane with the plane of the rotation path of the lineation parallels the slip lines. Depending upon the initial orientations of the lineation and the slip planes, the pole to the planar rotation path of the lineation may make any angle to the fold axis from 0°, where the lineation is perpendicular to the fold axis, to 90°, where the lineation parallels the fold axis, lies within the slip planes, and is therefore not rotated relative to the fold axis during slip (*Weiss*, 1955, pp. 228—229, 1959a, pp. 100—102; *Ramsay*, 1960, pp. 79—90).

These relationships recorded in the geometry of slip folds have been used to deduce the orientations of slip lines in about 50 folds in Trollheimen. For each fold, attitudes of the rotated lineation were measured on both limbs and at the hinge to orient the rotation path; these measurements were taken on a single folded surface within a given profile. Also measured for each fold was an attitude of the axial surface, which in a slip fold parallels the slip planes. The intersection of the axial surface with the plane of the rotation path was determined on an equal-area net, and its orientation was taken as the orientation of the slip lines for that fold.

The validity of this method of determining a slip-line orientation for any particular fold depends upon two conditions: (1) The lineation was straight before folding. (2) The fold was produced purely by slip. At present, neither of these conditions can be verified for any fold encountered in naturally deformed rocks. Nevertheless, where the lineation is seen to be straight within nonfolded fields surrounding the fold in question, probably it was also straight in the folded field. If the fold in profile is geometrically similar, probably it was produced by slip. However, the most practical criteria used in this study are based upon the nature of the rotation path of the lineation itself and upon its orientation relative to the fold axis. Where rotation of the lineation is large (greater than 90°) so that the difference between a cone and a plane is clear, and where the rotation path is planar and its pole not parallel (preferably 20° or more) to the fold axis, it is probable that a slip-line orientation for the fold under consideration is valid.

Interference Structures from Superposed Slip Folds

When an early set of folds is refolded by slip along planes transverse to the early fold axes, interference structures in the form of domes, saddles, and basins may result (*Reynolds* and *Holmes*, 1954, pp. 435—443; *Weiss*, 1959a, p. 97, Fig. 5; *Carey*, 1962, pp. 110 to 114; *O'Driscoll*, 1962, 1964; *Ramsay*, 1962b). These structures can be separated by shape into two general groups that reflect the orientation of the slip lines of the late set of folds relative to the limbs of the early set (Fig. 13; *Ramsay*, 1962b, p. 480, Figs. 14 and 15). Where the slip lines are oriented inside the interlimb angle of the early folds (group I, Fig. 13*a*), the profile sections[4] of the resultant domes and

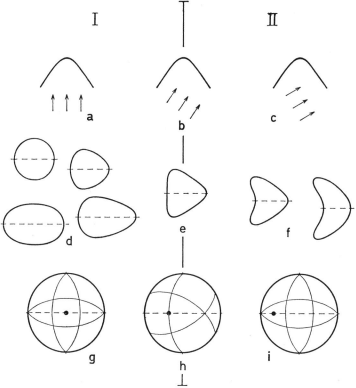

Fig. 13. Structural elements of interference domes and basins. *a, b,* and *c.* Relative orientations of late superposed slip lines (arrows) and sections of early folds; sections parallel the late slip planes. *d, e,* and *f.* Profiles of domes or basins produced from interfering folds oriented as in diagrams *a, b,* and *c,* respectively; dashed lines show traces of axial planes of the late slip folds. *g, h,* and *i.* Equal-area projections of attitudes of late axial planes (dashed great circles), layering (solid great circles), and late slip lines (dots) from the structures in diagrams *a* and *d, b,* and *e,* and *c* and *f,* respectively

[4] A profile of a fold is a section cut perpendicular to the fold axis (*Wegmann,* 1929, p. 107; *McIntyre,* 1950, p. 331). Although no single fold axis exists in a dome or basin, the term profile is still useful. Defined with respect to a regular upright dome with circular plan, a profile of the dome is a section cut perpendicular to its vertical axis of symmetry. With respect to a less regular dome or basin that lacks an axis of symmetry, a profile is a section cut perpendicular to its stacking axis, or direction of nesting of the layers involved.

basins range in shape from circles and ovals to triangles (Fig. 13d); where the slip lines are excluded by the limbs (group II, Fig. 13c), the profile sections range in shape from triangles to indented canoe shapes (group II, Fig. 13f). The special case that separates these two general groups occurs where the slip lines of the late folds parallel one of the limbs of the early folds (Fig. 13b); the resultant domes and basins are triangular in profile (Fig. 13e). These differences in shape provide a clear separation of such interference structures and are easily recognized in the field[5].

It follows from these relationships that, in an interference dome or basin that is triangular in profile section (Fig. 13e), the slip-line orientation of the late folds is contained by two planes; it parallels the axial surface of the late fold (which in turn parallels the late slip planes), and it parallels the early fold limb that remains planar through the late folding and that is transected by the late axial surface (Fig. 13b, e). Therefore, one need only measure the attitude of these two planes, find their intersection in spherical projection, and take the orientation of the intersection as the orientation of the late slip lines (Fig. 13b).

Similarly it follows that, in an interference dome or basin whose profile section falls into group I (Fig. 13d), the late slip lines parallel the axial surface of the late fold and are confined by the attitudes of the early fold limbs that are transected by the late axial surface (Fig. 13a, d). Therefore, one need only measure the attitude of these three planes, find from their intersections in spherical projection the angle within the axial surface included by the early fold limbs, and take that angle as the possible range of orientations of the late slip lines (Fig. 13g). Unfortunately, no unique slip-line orientation is obtained from this method unless the early fold is perfectly isoclinal; nevertheless, depending upon the form of the early fold, the slip-line orientation may be confined very closely (within $10°$).

Where it is difficult or impossible to measure an attitude of the late axial surface, certain other measurements can be substituted, and a solution for the orientation of the late slip lines can still be obtained. Let us consider the hypothetical fold represented in the lower-hemisphere, equal-area projection of Fig. 14a. Three attitudes (S_1, S_2, S_3) of the folded surface intersect at the fold axis, B; the axial surface (not shown) is vertical and confined by the two limbs, S_1 and S_2. Superimposed upon this fold is a second fold that formed by

---

[5] Although illustrated in Fig. 13 for interfering fold sets with perpendicular axial surfaces (cf. O'Driscoll, 1962, Fig. 2), these relationships hold also for the somewhat less regular shapes produced where the axial surfaces are not perpendicular (cf. O'Driscoll, 1962, Figs. 4, 5, 6).

slip along planes parallel with S'; the slip-line orientation is shown by an open circle. This situation is similar to that in Fig. 13*a*, and, assuming that the two folds interfere additively, a dome or basin will be produced that falls into group I by shape of its profile section (Fig. 13*d*). As the second fold develops by slip along S', B is rotated in a planar path toward parallelism with the slip lines (Fig. 14*b*;

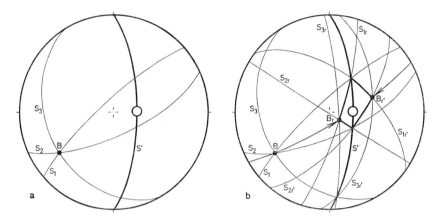

Fig. 14. Equal-area projection of early fold elements rotated during later slip folding to form an interference dome or basin. *a* Initial attitudes of the early fold axis, B, and the folded surface, S, superposed by late slip planes, S', and slip lines (open circle). *b* Early fold elements rotated in opposite directions (arrows) on opposite limbs (subscripts *r* and *r'*) of the late slip fold

Weiss, 1959a; *Ramsay,* 1960); B_r and $B_{r'}$ represent attitudes of B rotated in opposite directions on the two limbs of the fold. As the fold axis (B) is rotated toward the slip-line orientation, the folded surface (S_1, S_2, S_3) is rotated toward parallelism with the slip planes (S'), but only with infinite slip would B parallel the slip lines and S the slip planes. Therefore, with some finite amount of slip, the attitudes of B and the folded surface, S, approach the orientation of the slip lines but never parallel them.

This relationship permits us to limit the orientation of the late slip lines of an interference dome or basin in group I by the attitudes of its limbs alone. It was found practical in Trollheimen to measure four or more attitudes of a folded surface around such a dome or basin, plot them as great circles on an equal-area projection, and take their polygon of intersections (emphasized by heavy lines in Fig. 14*b*) as the possible range in orientation of the late slip lines. Identification of the polygon that contains the slip-line orientation depends upon noting where the center of the closure lies relative to each plane and finding the polygon in spherical projection that satisfies this relationship for all the planes.

A dome or basin whose profile section falls into group II (Fig. 13f) cannot be used in this way to determine the slip-line orientation of the late fold. Although the late slip lines parallel the late axial surface, they are oriented outside the interlimb angle of the early fold and therefore are not closely confined (Fig. 13c, i).

The validity of this method of using attitudes of the limbs and axial surface to deduce the slip-line orientation for a dome or basin depends upon three conditions: (1) The dome or basin is the product of two interfering folds. (2) One fold can be identified as early and the other late. (3) The late fold was produced by slip. Proof that the first two conditions hold for any particular dome or basin depends upon the individual geologic situation. The two criteria found most convincing with respect to such structures in Trollheimen are the patterns made by whole groups of them and the presence of an axial-surface schistosity. Domes, basins, and saddles in Trollheimen occur at points of intersection in a two dimensional, more or less right-angle grid. Throughout the grid, a dome is connected to another dome by a saddle along a grid line; a basin is connected to another basin by a saddle along a grid line; domes and basins alternate along diagonals. This pattern is characteristic of two interfering sets of folds (O'Driscoll, 1962, 1964; Ramsay, 1962b, p. 468, Fig. 2). An undeformed pervasive schistosity, which parallels the axial surfaces of isoclinal folds throughout Trollheimen, parallels the axial surfaces of the domes and basins and contains one set of the grid lines. This indicates that the grid lines contained by the schistosity represent the hinge lines of a late set of folds.

Verification of the third condition that the late fold was produced by slip relies in good part upon the criteria discussed in the foregoing section on slip folding of lineations. An additional consideration, however, is the geometrical impossibility of forming tight interference structures by late flexural-slip folding without creating voids between the layers or without developing subordinate folds and faults. Therefore, if the dome or basin is simple (without subordinate structures) and tight, so that the polygon of intersections defined by attitudes of its limbs is small enough to result in a useful, limited range in slip-line orientations, the late fold was almost certainly produced by slip.

Asymmetry and the Separation Angle

Groups of folds commonly display planar preferred orientations of fold axes. Study of the asymmetry of such folds in Trollheimen has led to the concept of the *separation angle,* which can be defined as the planar angle that separates fold axes by orientation into groups of opposite asymmetry. Folds in many different structural situations may display a separation angle, the kinematic significance of which depends upon the initial geometry and the mechanism of folding.

Flexural-Slip Folds in Planar Layers

The following example serves to illustrate a single generation of flexures that display a planar preferred orientation of fold axes and to introduce the concept of the separation angle.

Tundra Landslide. The northwest slope of Blåhø in Trollheimen is dotted with small landslides in the tundra sod. A map of one such landslide, approximately 6 meters in diameter, is shown in Fig. 15*a*.

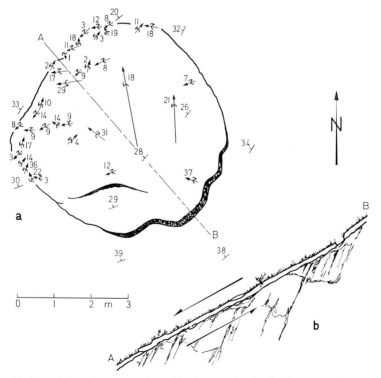

Fig. 15. Map *(a)* and cross section *(b)* of a tundra landslide on northwestern Blåhø, Trollheimen, Norway. Horizontal and vertical scales equal

3 Hansen, Strain Facies

64856

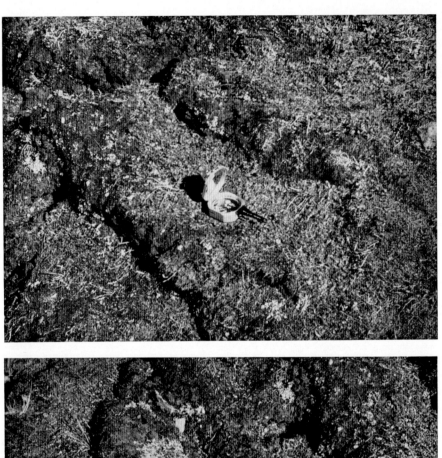

Pl. 2. Folds in tundra sod within a landslide on northwestern Blåhø, Trollheimen, Norway. Brunton compass indicates size

Strike-and-dip symbols around the landslide indicate the slope of the tundra surface where unaffected by the slide; their dips average 33° to the northwest, and their strikes indicate that the landslide is located in a slight indentation in the slope. The scar at the southeastern end is a gap in the sod, less than 30 centimeters across, where the underlying soil and bedrock are exposed, and where the thickness of the landslide measures 15 centimeters. Elsewhere, along the sides and within interior tears, the slide shows about the same thickness.

The remaining structural symbols on the map represent folds in the tundra sod. Linear arrows show the orientations of fold axes and the relative lengths of hinge lines, and semicircular arrows modifying the axes indicate their asymmetry in the downplunge direction. The folds are parallel in profile, the sod appearing to have flexed like a heavy rug (Pl. 2). Their height-width ratios are small, all less than 0.5, and their depth-width ratios are also small because the folds are disharmonic on the sod-bedrock interface. Most of their hinge lines are short and curved, and on the scale of cylindricity the folds are cylindroidal and irregular.

A good part of the movement history of this landslide can be assumed. The tundra sod has moved downslope under the influence of gravity and has overridden the bedrock. The surface of slip, the sod-bedrock interface, has about the same attitude as the tundra's upper surface because the slide everywhere has the same thickness.

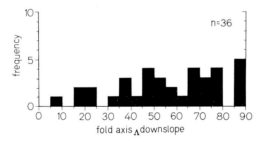

Fig. 16. Histogram of angles between fold axes and the mean downslope direction in the tundra landslide

The mean downslope direction, obtained by averaging the downslope directions of the tundra surface both within and around the slide, trends N42°W and plunges 31°NW. This downslope movement line, or slip line, parallels the shear couple acting along the slip plane, the overlying sod having moved downward and northwestward relative to the underlying soil and rock. The dashed line *AB* in Fig. 15*a* and the cross section *AB* in Fig. 15*b* parallel the slip line.

3*

The histogram in Fig. 16 shows the angle between each fold axis in the landslide and the mean downslope slip line for the whole slide. It is noteworthy that no strong mode is present; fold axes appear at practically all angles to the mean slip line, though only 10 of the 36 axes are oriented at angles of 45° or less.

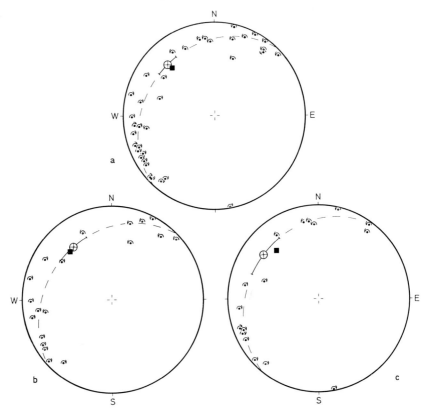

Fig. 17. Fold axes (dots) and asymmetry (semicircular arrows) of folds in the tundra landslide. Separation angles (arcs with circular symbols) are shown in planes approximating the fold-axis distribution (dashed great circles). The mean downslope direction is shown by squares. *a* Composite diagram, 36 axes, 18° separation angle. *b* Partial diagram of folds southwest of line *AB*, Fig. 15*a*; 19 axes, 27° separation angle. *c* Partial diagram of folds northeast of line *AB*, Fig. 15*a*; 17 axes, 34° separation angle

Fig. 17*a* shows the orientations of the slip line (square) and the 36 fold axes in a lower hemisphere, equal-area projection. The fold axes are distributed along a great-circle girdle (dashed line) which is oriented within a few degrees of the average position of the tundra surface and therefore essentially parallels the surface of slip for the landslide. When the asymmetry is examined, it becomes apparent that all the axes displaying clockwise asymmetry plunge toward the west, and all but one of the counterclockwise axes plunge toward the north. Separating the two groups of axes — the clockwise from the counterclockwise — is an angle which measures 18° within the plane of preferred orientation of axes and which is shown as an arc of the great circle in the projection. The downslope direction — the slip line for the landslide — is oriented between these groups of axes with opposite asymmetry, within the separation angle.

The circular symbol drawn within the separation angle in Fig. 17a represents the tail of an arrow. It indicates that the upper component of the shear couple compatible with the asymmetry of the folds in the landslide is directed downward, away from the reader, parallel with the slip surface. If the axes with northward plunges were clockwise and those with westward plunges counterclockwise, they would indicate an upper shear component directed upslope, toward the reader, and the circular symbol would contain a dot, rather than a cross, and would represent the head of an arrow.

Projections b and c in Fig. 17 show the fold axes in the landslide divided — those to the southwest of AB are plotted in b, and those to the northeast of AB are plotted in c; there are 19 axes in b and 17 in c. In both diagrams the axes describe a girdle distribution, as in diagram a, though their numbers are depleted by half. Likewise in both diagrams the axes are grouped by asymmetry, and the separation angles contain the slip-line orientation. The only essential difference between the composite projection with all 36 axes and the two local projections is the size of the separation angle. In the local projections the angles are 27° and 34°, and in the composite projection the angle is narrowed to 18°. These two partial diagrams emphasize the fact that the westward-plunging clockwise folds are not located primarily on the northeastern side of the landslide, and the northward-plunging counterclockwise folds are not primarily on the southwestern side, as might be expected intuitively. Instead, both kinds of axes are shared evenly by both sides of the slide, and the separation of axes by asymmetry and orientation is not coordinated with a separation by location. Therefore, the separation angle in each of the local projections confines the slip-line orientation of the landslide almost as well as the angle in the composite projection.

Comparison of the observed fabric of the flexures in the landslide with the independently known movement history of the slide has shown the following relationships: Folds have formed with axes oriented at all angles, from perpendicular to nearly parallel, to the

slip-line orientation for the landslide. The planar preferred orientation of fold axes parallels the slip surface of the slide. The separation
angle contains the slip-line orientation of the slide. The distribution
of axes according to asymmetry is compatible with the shear couple
or slip direction for the domain as a whole. Finally, partial diagrams,
such as might be produced by a geologist working in an area of limited exposure, can represent accurately the fabric and movement
history of the whole domain.

Had the tundra sod been composed of two or more layers instead
of one, slip would have occurred between the layers during flexing,
and, strictly speaking, the folds would be flexural-slip folds instead
of flexural folds. Nevertheless, the kinematic relationships concluded
in the previous paragraph are not changed by the addition of layering
to the folding mass, and therefore they apply as well to the more
general case of flexural-slip folds.

General Analysis. It is useful to reconstruct some of the relationships between stress and strain that obtained within the deforming
landslide. As the following discussion shows, these relationships need
only be known generally for our present purpose, and so the details
of the dynamics that would not alter the conclusions on the kinematics are ignored. Throughout this discussion mechanical isotropy is
assumed for the sod.

The stress field due to gravity under which the tundra sod began
to move downslope was triaxial ($\sigma_1 > \sigma_2 > \sigma_3$) with σ_1 vertical; σ_2 was
parallel with strike because the indentation in the slope caused the
two sides of the landslide to tend to converge. The parallel fold
geometry is permissive evidence that the sod was not passive during
deformation but was strong enough to refract the stress field (dotted
lines, Fig. 18a) in such a way that σ_1 would tend to shift from the
vertical toward parallelism with the downslope direction (one possible orientation arbitrarily designated). The resolved shear stress
on the sod-bedrock interface *(S)* would attain a maximum value at
the intersection of the plane of σ_1 and σ_3 with the interface, i. e. parallel with the line of dip, in the sense of top downward. Assuming
that the landslide can slip with equal ease in all directions along the
bedrock surface, the slip line of the landslide would parallel the
orientation of the maximum resolved shear stress, and slip would
occur in the sense of the shear couple.

The primary principal strain axes would be approximately par-
allel with the stress axes at initial strain, so that $\varepsilon_1 \| \sigma_3$, $\varepsilon_2 \| \sigma_2$, and
$\varepsilon_3 \| \sigma_1$. The landslide displays no evidence of rotation about the pole
to the slip surface ($\perp S$) during its short translation, and it may therefore be assumed that ε_2 did not rotate out of parallelism with σ_2.

Axial planes $(S'_1, S'_2, \ldots, S'_6)$ of the folds would contain the fold axes $(b_1, b_2, \ldots, b_6)$ by definition, as well as the line of easiest release $(\sigma_3, \varepsilon_1)$, and therefore would be cozonal, the zone axis being parallel with ε_1. By the relationship shown in Fig. 10, the senses of the dashed arrows drawn in Fig. 18a from the pole to the mean orientation of the tundra surface $(\perp S)$ to the poles to the axial planes $(\perp S'_1, \perp S'_2, \ldots, \perp S'_6)$ identify the asymmetry of the folds, shown by the semicircular arrows modifying their axes. The asymmetry of

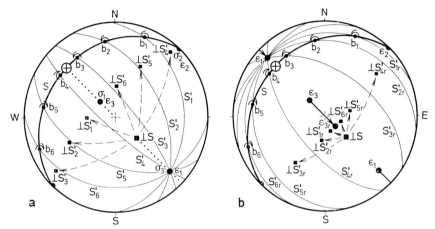

Fig. 18. Hypothetical relationship of inferred principal axes of stress and strain to the structural elements in the tundra landslide. The orientation of the maximum resolved shear stress in shown by the circular symbols in the slip planes, S. The 36 fold axes in the landslide are represented schematically by axes $b_1, b_2, \ldots, b_6$.
a Initial geometry. *b* Geometry after rotation due to drag

all the folds is compatible with the sense of the shear couple deduced from the inferred stress field, and the separation angle (between b_3 and b_4) confines the orientation of the maximum resolved shear stress, as well as the slip line, on the slip plane. These relations do not depend on the amount of refraction of the stress field.

As the landslide moved downslope, the friction at the sod-bedrock interface would have caused the part of the sod at its lower surface to move less freely than the rest of the sod above it. If the landslide were composed of several layers rather than one, the effect of this drag might have been considerably more pronounced, but nevertheless its possible geometrical consequences must be considered. The principal strain axes ε_1 and ε_3 would have tended to rotate out of parallelism with σ_3 and σ_1 toward parallelism with the slip line and pole to the slip plane, respectively (e. g., ε_{1r}, ε_{3r}, Fig. 18b); rotation would have occurred about the horizontal line ($\|\sigma_2$, ε_2) perpendicular to the slip line, in the sense of the shear couple. During this process, the orientations of the fold axes would not have changed unless the slip varied appreciably in direction and/or magnitude within the landslide, but the axial planes would have rotated toward parallelism with the slip plane, their zone axis remaining parallel with ε_{1r}. Upon extreme rotation, the axial planes could have approached, but never passed, parallelism with the orientation of the tundra surface, though of course the whole landslide could have tumbled (somersaulted) through several complete circles relative to the stress field. The dashed arrows in Fig. 18b show that rotation due to this kind of drag changes neither the asymmetry of the folds nor the size or orientation of the separation angle.

If folds were to form semichronologically while the sod was undergoing rotation due to drag, each sucessive axial plane as it came into being would initially parallel ε_1 but would immediately be rotated toward ε_{1r} and the slip line. The amount of rotation of the axial planes would differ from fold to fold and, ideally, would depend upon when they began to develop within the process of rotation; those forming earlier in the process would tend to have been rotated more than those forming later. In this situation, therefore, the axial planes of the folds would not share a zone axis (ε_{1r}) but in spherical projection would intersect the angle between ε_1 and the slip line at different points.

Had a fold developed in the landslide with its axis parallel with the slip line, its axial plane would be vertical and perpendicular to the mean orientation of the surface of the sod, and the fold would be symmetrical (Fig. 10b). Theoretically, therefore, the separation angle of a group of folds with a planar preferred orientation of axes reduces to the orientation of a line parallel to which the axes of symmetrical folds develop but to either side of which the axes of asymmetrical folds develop with opposite asymmetry. Therefore, the term *separation line* is used in the following theoretical discussion with the understanding that it corresponds to the separation angle encountered in natural folds, in which it is rarely reduced to a line.

A more general relationship between the principal axes of stress and strain and the separation angle of flexural-slip folds is illustrated in Fig. 19. Its only restriction is that no rotation has occurred about the pole to the slip plane. None of the principal stress axes parallels the layering (S) which acts as the slip plane, and, in a triaxial stress field as shown, the orientation at which the resolved shear stress attains its maximum value on the slip plane departs from the plane of σ_1 and σ_3 (dashed great circle) by a variable amount determined by

the relative values of the principal stresses and by the orientation of
S relative to them. An arbitrary orientation of the maximum resolved
shear stress is shown by the circular symbol on the slip plane, and,
assuming slip with equal ease in all directions within the slip plane, it
parallels the slip line. In the diagram, the principal strain axes are

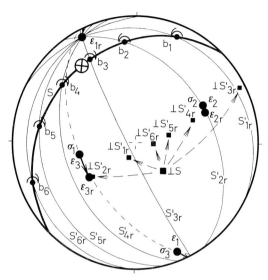

Fig. 19. General relationship between the principal axes of stress and strain and
the separation angle of flexural-slip folds. The orientation of the maximum
resolved shear stress is shown by the circular symbol in the slip plane, S

shown displaced (solid arrows) from the principal stress axes by rotation due to drag. The separation line of the folds parallels the intersection of the planar layering (S) with the plane that contains ε_{1r} and
the pole to the layering ($\perp S$), but it does *not* parallel the slip line. It
is possible, though, with the proper values for the principal stresses, the
correct orientation of the slip plane, and the right amount of rotation
due to drag, that the separation line may fortuitously parallel the slip
line. However, if σ_3 were refracted to parallel the slip plane (S), the
situation would become the same as that shown in Figure 18*b*, and
the separation line would parallel the slip line. In conclusion, therefore, if no rotation occurs about the pole to the slip plane, the separa-
tion line of flexural-slip folds parallels the slip line when ε_2 ($\parallel \sigma_2$)
parallels the layering, but, except in special cases, it does not parallel
the slip line when ε_2 ($\parallel \sigma_2$) does not parallel the layering.

Unfortunately this criterion is not dependent upon the geometry
of the deformed mass alone and consequently is difficult to use to

assess the kinematic significance of a separation angle defined by flexural-slip folds encountered in the field. Nevertheless, the meager amount of data on the orientation of stress fields recorded in rocks shows that σ_2 parallels the layering in rocks folded by the flexural-slip mechanism (*McIntyre* and *Turner*, 1953; *Gilmour* and *Carman*, 1954; *Clark*, 1954; *Hansen* and *Borg*, 1962; *Scott, Hansen,* and *Twiss*, 1965; *Carter* and *Friedman*, 1965). Although these certainly may constitute a fortuitous and nonrepresentative sample, they indicate that, in rocks in which the layering is strong enough to cause folding by flexural slip, it is also strong enough to refract the stress field so that σ_2 becomes practically parallel with the layering. At least, this is permissive evidence that, for any group of natural flexural-slip folds (including the flexures in the tundra landslide), σ_2 was parallel with the layering, and the separation angle confines the slip-line orientation.

Groups of flexural-slip folds displaying wide variations in axial orientations and displaying asymmetry relationships consistent with unique slip directions have been described by *Martin* (1961, pp. 19 to 27), *Hansen, Porter, Hall,* and *Hills* (1961), *Page* (1963, p. 668, Fig. 5), *Howard* (1968), *Scott* (1969), *Scott* and *Hansen* (1969), *Hansen* and *Scott* (1969), and *Stanley* (1969, p. 51, Fig. 3).

Slip Folds in Planar Layers

It is customary to think of the slip that occurs during slip folding as that which takes place between particles on opposite sides of the slip planes. A slip fold produced in planar layers by only this type of slip would be unchanging in profile along a straight hinge line (Fig. 20a). In order for slip folds to have hinge lines with finite dimensions, however, a certain amount of slip must occur between particles within the slip planes themselves. A slip fold produced in planar layers by both these components of slip would vary in profile along a curved hinge line (Fig. 20b, c).

During formation of such a fold by both components of slip, the hinge line is rotated in a planar path, parallel with the slip planes, toward perallelism with the slip lines. This model assumes the simple case (and probably the most common geologically) of the intersection of the slip planes with the compositional layering localizing the hinge line, so that, as the fold develops and the hinge line rotates, it cannot rotate out of parallelism with the slip planes because there is no component of slip across those planes. By the same argument, the hinge line cannot be rotated beyond parallelism with the slip lines because no component of slip exists across those lines. Therefore the maximum bend that a hinge line may attain is 180°, which can happen only by infinite slip between adjacent particles within the slip planes; such a hinge line would be U-shaped with both prongs of the U parallel with the slip lines. It follows from these relationships that attitudes of the fold axis measured at different positions along the curved hinge line define a plane which parallels the slip planes and that the continuum of axial attitudes must exclude the slip lines (Fig. 20c).

In a slip fold that develops in planar layers, the attitude of the axial plane, which parallels the slip planes, and the attitude of the

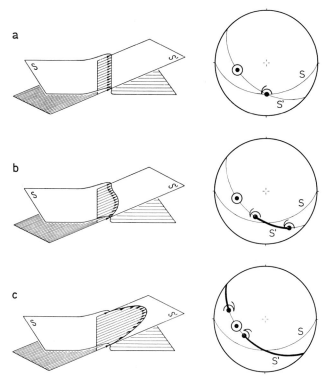

Fig. 20. Folds with straight and curved hinge lines produced in planes, S, by slip (parallel with arrows) within intersecting planes, S'. In projection, slip-line orientations are shown by circular symbols, fold axes by dots, and continua of axial orientations by heavy great-circle arcs; patterns of asymmetry are shown by semicircular arrows

layers surrounding the fold are constant along the length of its hinge line, whether the hinge line is straight or curved. Therefore, the angle between these two planes and their orientations relative to each other are also constant along the hinge line; because this relationship governs the asymmetry of a fold (Fig. 10), the asymmetry must also be constant along the hinge line[6]. Consequently such a fold, by its axial attitudes and asymmetry, defines a separation angle which measures $180°$ if its hinge line is straight and measures from $180°$ to $0°$ if its hinge line is curved.

If a group of such folds formed and their fold axes were measured upon the exposed portions of their hinge lines, the resultant preferred orientation of axes would be planar, parallel with the slip planes, and a separation angle would be defined that contains the orientation of the slip lines.

Superposed Slip Folds of a Single Order; the Hinge-Line Node

Consider now the situation in which a set of slip folds develops only by slip between the slip planes, with no component of slip within the planes. If the compositional layering is relatively planar, the resulting folds display a linear preferred orientation of fold axes, but if the layering is nonplanar, as in a previously folded field, the resulting superposed folds may display a planar preferred orientation of axes. The following discussion deals with the relationship between separation lines and slip lines of such superposed folds with planar distributions of axes.

The isometric block diagram in Fig. 21a shows part of a folded layer; the upper surface of the block is horizontal. Several attitudes of the folded layer $(S_1, S_2, \ldots S_9)$ intersecting at the fold axis (B) are shown projected to the horizontal plane in the lower-hemisphere equal-area projection of diagram g. Block b shows block a folded by differential slip along shallow slip lines within steeply dipping planes. The orthogonal profiles in diagram c show the direction and relative amount of displacement of the outer surface of the layer in block a to form the superposed folds in block b. Attitudes of the slip planes (S') and slip lines (open circle), as well as several of the superposed fold axes $(b'_1, b'_2, \ldots, b'_9)$ at the intersections of S' with the folded layer, are shown in diagram g. Dashed arrows are drawn from the poles to the layer $(\perp S_1, \perp S_2, \ldots, \perp S_9)$ toward the pole to the slip planes $(\perp S')$; following the relationship described with reference to Fig. 10, the dashed arrows indicate the asymmetry of the super-

---

[6] An apparent exception to this statement is found in a fold with a curved hinge line that passes through the horizontal plane, because at that point, by the convention of describing a fold's asymmetry by looking down the plunge of its hinge line, we change from looking toward one end of the hinge line to looking toward the opposite end, which of course displays the opposite asymmetry (Fig. 20c).

posed folds, as shown by semicircular arrows modifying their axes $(b'_1, b'_2, \ldots, b'_9)$.

The superposed fold axes that plunge toward the west are clockwise, and those that plunge toward the south are counterclockwise;

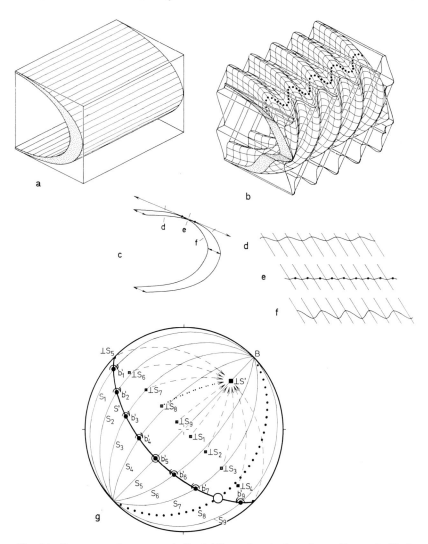

Fig. 21. Geometry of superposed slip folding of a single order. *a* Isometric block diagram containing part of a folded layer. *b* Block *a* folded into slip folds of a single order. *c* Profiles of the outer folded surface in block *a*. *d*, *e*, and *f*. Successive profiles of the superposed slip folds in block *b*. *g* Lower-hemisphere equal-area projection of structural elements in blocks *a* and *b*

b'_5, between these two groups, is symmetrical (Fig. 21g). Therefore, a separation line exists parallel with b'_5, and it also parallels the intersection of the slip planes (S') with the plane that contains the earlier fold axis (B) and the pole to the slip planes ($\perp S'$). Nevertheless, the separation line does not parallel the slip-line orientation, which is $40°$ away. If the earlier fold axis were oriented fortuitously so as to parallel the plane that contains the slip lines and the pole to the slip planes, only then would the separation line parallel the slip lines. In conclusion, therefore, the separation line defined by such superposed slip folds does not parallel the slip lines, except in special cases.

Diagrams d, e, and f in Fig. 21 are successive profile sections of the outer surface showing the superposed folds. The viewer is looking down the plunge of the superposed fold axes, opposite to the direction in which the folds are seen in block b. The axial surfaces are continuous through all three sections; they are shown broken only because the hinge lines are curved and the profiles are consequently taken in different planes. Profile d has been constructed approximately perpendicular to b'_9, profile e perpendicular to the slip-line orientation, and profile f perpendicular to b'_7 (Fig. 21g); their locations and orientations are indicated in diagram c. The folds in both diagrams d and f, with axes oriented on opposite sides of the slip-line orientation, show counterclockwise asymmetry, thereby illustrating the fact that in this case the slip-line orientation does not separate axial orientations of opposite asymmetry.

Every trough in the folded surface shown in profile d is aligned along its axial surface with a crest in profile f, and every crest is aligned with a trough. Every hinge line of the superposed folds in block b changes from a crest to a trough along the hinge line. The point at which the form of the fold changes from convex to concave along the hinge line is called the *hinge-line node*. At the node, the folded surface is tangent to the slip lines, and no folding occurs (Fig. 21c). The dotted line in block b is the locus of points (including nodes) at which the folded surface is tangent to the slip lines; the dots in profile e are its intersections with the profile section; and the dotted great circle (S_8) in diagram g shows its attitude. The intersection of the axial surface of one of the superposed folds with the plane tangent to the folded surface at the hinge-line node parallels the slip lines. Therefore, in such a system of superposed slip folds, favorably exposed, the hinge-line node rather than the separation line can be used to deduce the slip-line orientation.

Slip between particles within the slip planes does not change the orientations of the slip lines, the slip planes, or the early fold axis

(*B*) surrounding the late folds. Consequently the orientations of the separation line and the surface (*S*) that contains the hinge-line node are independent of slip within the slip planes, and the conclusions drawn in the foregoing paragraphs need not be altered to include the more general case of slip folding by both kinds of slip.

Superposed Slip Folds of Multiple Orders

Let us now consider what happens to the shear senses of slip folds and their separation line when the surfaces that display the folds (e.g., bedding) are rotated toward parallelism with the slip planes.

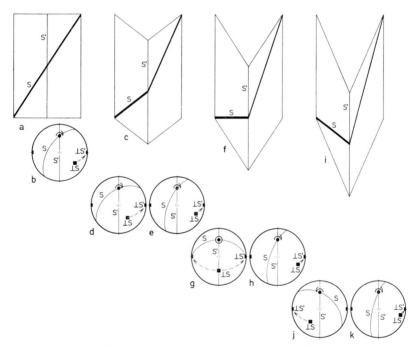

Fig. 22. Rotation of layering, *S*, relative to the slip planes, *S'*, during progressive development of a low-order slip fold. Projections show the orientations of elements that govern the asymmetry (semicircular arrows) of high-order folds on the limbs of the low-order fold

Such rotation occurs, for example, when two or more orders of slip folds develop contemporaneously. The schematic cross sections in Fig. 22 illustrate the progressive development of a slip fold with higher-order slip folds on its limbs; the elements that determine the asymmetry of the higher-order folds are shown in separate equal-area projections for each limb of the lower-order fold in the successive cross sections. In the example illustrated, the slip planes make such a low angle with the layering before folding is initiated (section a) that, as the low-order fold begins to form and the layering rotates slightly (section c), any high-order folds that form will have the same asymmetry on both limbs (clockwise, diagrams d, e); with respect to the lower-order fold, they are reverse on the left limb and normal on the right. As the low-order fold continues to develop, the trend of the layering on the left limb rotates until it becomes perpendicular to the slip planes (section f), and the high-order folds on that limb become symmetrical (diagram g); the high-order folds on the right limb remain asymmetrical and clockwise (diagram h). As the low-order fold develops further, the trend of the layering on the left limb rotates past its perpendicular position (section i), and the high-order folds become asymmetrical again but counterclockwise (diagram j); with respect to the lower-order fold, the high-order folds on both limbs are now normal (diagrams j, k). This example is not intended to be a hypothesis for the manner of formation of multiple orders of slip folds but merely an illustration of the geometric consequence of rotation of the layering upon their asymmetry.

If the pattern of slip that produced the two orders of folds drawn in Fig. 22 were imposed upon the recumbent fold shown in Fig. 21a, the early fold axis (B) would rotate on the limbs of the superposed low-order fold toward parallelism with the slip lines ($Weiss$, 1969a; $Ramsay$, 1960). Two positions of the rotated fold axis (B_r, $B_{r'}$) on the right limb of the low-order fold are shown in the projection in Fig. 23a; the same orientations of slip lines and slip planes that resulted in the single order of superposed folds in Fig. 21b are used here. As B is rotated toward the slip-line orientation, the plane containing B and the pole to the slip planes ($\perp S'$) also rotates toward the slip-line orientation; consecutive attitudes of this plane ($B \cdot \perp S'$, $B_r \cdot \perp S'$, $B_{r'} \cdot \perp S'$) appear as dashed great circles in Fig. 23a. It was shown in Fig. 21g that the intersection of this plane ($B \cdot \perp S'$) with the slip planes (S') parallels the separation line. Therefore as $B \cdot \perp S'$ rotates toward the slip-line orientation through $B_r \cdot \perp S'$ and $B_{r'} \cdot \perp S'$, the separation line of the high-order folds must rotate from b_5' through b_6' and b_7' toward the slip-line orientation (Fig. 23a). However, only as the amount of slip becomes infinitely large does the separation line approach parallelism with the slip lines. Fig. 23b shows rotation of these elements on the left limb of the superposed low-order fold; rotation occurs in the opposite sense from that on the right limb (diagram a). Another case of the rotation of all the elements that determine asymmetry and the separation line is illustrated in detail in Fig. 23c.

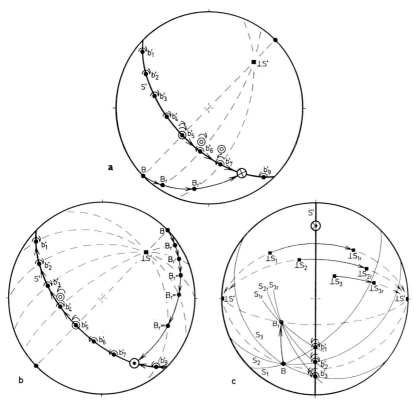

Fig. 23. Rotation of the structural elements that determine the separation line in superposed slip folds. *a* Elements in Fig. 21*b* progressively rotated on the right limb of the low-order fold in Fig. 22. *b* The same elements rotated in the opposite sense on the left limb of the same low-order fold. *c* Another special case showing the rotation of the separation line

Where an early fold is superposed by two orders of slip folds, the angle Δ between the late slip lines and the separation line of the late high-order folds is a trigonometric function of three independent variables (Fig. 24): (1) the angle Φ between the late slip planes (S') and the mean orientation (or trend) of the layering (S_{tr}) on the limb of the late low-order fold, measured between their poles, (2) the angle θ between S' and the rotation path of the early fold axis (B_{path}), measured between their poles, and (3) the angle τ between the late slip lines and the late low-order fold axis (B'). From considerations of the geometry, it can be shown that

$$\tan \Delta = \frac{\sin \tau \tan \Phi}{\tan \theta + \cos \tau \tan \Phi}$$

where $0° < \Phi < 180°$, $0° < \theta \leq 90°$, and $0° < \tau < 180°$.

4 Hansen, Strain Facies

Clearly, if we could determine the value of these three variables in any given locality, we could determine the angle between the slip lines and the separation line simply by solving this equation. Unfortunately, we cannot know the value of τ without knowing the orientation of the late slip lines, which in fact we are trying to find.

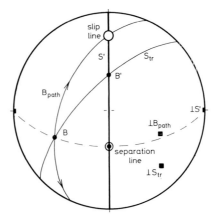

Fig. 24. Equal-area projection of structural elements that determine the orientation of the separation line of late high-order folds relative to the late slip lines of superposed slip folds of two orders

However, assuming that all values for this angle are possible, we can derive an equation for the maximum angle (Δ_{max}) between the slip lines and the separation line by solving simultaneously the above equation and its partial derivative with respect to τ, eliminating τ between them:

$$\tan \Delta_{max} = \frac{\tan \Phi}{(\tan^2 \theta - \tan^2 \Phi)^{1/2}}$$

In order to determine the value of θ we must know the orientation of the rotation path of B; but, knowing this, we can solve directly for the orientation of the late slip lines by finding the intersection of the late slip planes (S') with the rotation path (*Weiss*, 1959a; *Ramsay*, 1960), and not having to bother with the separation line at all. Nevertheless, use of the separation line to find a slip-line orientation by the foregoing equations can serve as a check on the solution for the slip-line orientation from other methods.

Use of the Separation Angle

In order to use the separation angle to deduce the orientation of slip lines, it is necessary to determine that certain critical conditions are met: (1) the folds belong to a single generation; (2) they belong to a single order; (3) they are located between two adjacent axial surfaces of the next lower-order folds of the same generation. These three statements can be verified by inspection, depending upon the geologic situation, exposure, etc. Once these conditions are known to hold for any group of folds in question, it follows from theoretical considerations that the planar distribution of fold axes parallels the slip planes, regardless of folding mechanism or initial geometry.

It then becomes necessary to determine (4) whether the layers that display the folds were planar or nonplanar before folding. If the layers were planar before folding, if for example the folds are not superimposed upon an earlier set, or, if they are, they are studied within planar limbs of the earlier set, the separation angle probably confines the orientation of the slip lines. If the folds share a common attitude of axial planes that parallels the planar distribution of fold axes, geometrically they are slip folds, and their separation angle confines the slip-line orientation. If the axial planes of the folds are not mutually parallel, and the planar distribution of fold axes parallels the layering, geometrically they are flexural-slip folds, and their separation angle probably, though not necessarily, confines the slip-line orientation.

However, if the folds in question are involved with the geometry of another set of folds, it is necessary that (5) the folds from which the separation angle is obtained represent the latest generation of folds. Nevertheless, even if they do, the separation angle may not confine the orientation of the slip lines, but the relationship discussed at the end of the foregoing section can be used to assess the kinematic significance of such a separation angle.

It has been assumed in this discussion that the slip lines operating during the formation of the folds from which the separation angle is obtained were parallel throughout all the folds studied. Deviation from parallelism from fold to fold, or even within individual folds, would cause a certain amount of overlap of axial orientations of opposite asymmetry and cause a consequent increase in the size of the separation angle. Such a separation angle containing axial orientations of both clockwise and counterclockwise asymmetry might roughly measure the spread of orientations of the slip lines themselves.

IV. Geometric Types of Flow

It is known from stratigraphic displacements measured in tens of kilometers in some areas that macroscopic structures like nappes and domes in mountain chains result from the transport and flow of large volumes of rock. Mesoscopic structures displaying the same style, symmetry, and orientation of principal fabric axes as the macroscopic structures in which they are found indicate that the large-scale flow was penetrative to the mesoscopic scale. It follows that these mesoscopic structures record on their scale the same flow that is recorded on the macroscopic scale by the nappes and domes. It would appear, therefore, that an immediate goal in analyzing mesoscopic strain features should be the flow that they record. Once the flow recorded by separate generations of structures within the outcrops of a given macroscopic structure is sufficiently understood, it should be possible to integrate over that volume to arrive at the flow that occurred during the various phases of development of the structure in its different parts. By this procedure it may be possible to learn some significant facts about the processes in which such large-scale structures develop.

With this in mind, structures are analyzed here in terms of simple, geometric types of flow. In the discussion that follows, rock is considered an incompressible fluid, with the assumption that all density changes, which might occur by phase transformations or loss of volatiles, are negligible.

Definitions

A *stream line* in a fluid is a line that is tangent to the velocity vector at every point of the fluid's length; it shows the instantaneous direction of flow at every point (cf. *Rouse*, 1946). In Fig. 25, stream lines are drawn parallel with f, and axes d and e are drawn perpendicular to f at every point. These three orthogonal, kinematic axes may curve continuously throughout a fluid.

Diagrams a through k in Fig. 25 represent *stream tubes* generated by stream lines. The stream tubes are divided into smaller volumes that approximate cubes at their front faces. Each of the subvolumes has been constructed equal to every other subvolume in the same tube (and in every other stream tube as well, except tube k) and therefore their different shapes along the length of a tube represent the changing shape of a given volume of incompressible fluid as it proceeds through the tube. (Thus they are somewhat analogous to the changing shapes of the strain ellipsoid during progressive deformation, as discussed by *Flinn*, 1962.)

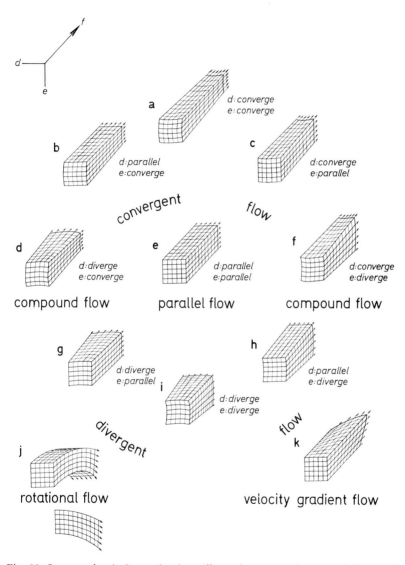

Fig. 25. Stream tubes in isometric views illustrating geometric types of flow, with reference to three orthogonal flow axes, *d, e,* and *f*. The cross section of tube *j* (bottom, left) parallels *ef*

The nine different stream tubes drawn in the ring were obtained by combining in every possible way stream lines that converge, remain parallel, and diverge with respect to the *d* and *e* axes. In *convergent flow*, illustrated by tubes *a*, *b*, and *c*, the stream lines either converge with respect to both axes or converge with respect to one and remain parallel with respect to the other; there is a decrease in cross-sectional area through which flow proceeds. In *divergent flow*, tubes *g*, *h*, and *i*, the stream lines either diverge with respect to both axes or diverge with respect to one and remain parallel with respect to the other; there is an increase in cross-sectional area. In *parallel flow*, tube *e*, the stream lines remain parallel with respect to both axes, and the area is unchanging. In *compound flow*, tubes *d* and *f*, the stream lines both converge and diverge, and the cross-sectional area may decrease, maintain the same size, or increase.

Convergent flow of an incompressible fluid, in the special case of *steady flow*, in which the velocity of the fluid at any given point in space is constant with time, can be termed acceleration flow; velocity increases in magnitude along a given stream line. Similarly, divergent and parallel flow can be termed deceleration and uniform flow, respectively (*Mackin*, 1947, pp. 26—28). However, these terms are not used in this book because they are not always descriptively accurate in the general case of *unsteady flow*. Any fluid, whether compressible or incompressible, undergoing unsteady flow, whether convergent, parallel, or divergent, may accelerate, maintain the same velocity, or decelerate along a given stream line.

Rotational flow occurs when a fluid takes a curved path. The stream lines are curves, and the angular velocity about every particle within the fluid is greater than zero. The fluid represented by the parallel stream tube in diagram *j* is undergoing rotational flow about the *d* axis.

When a fluid is flowing faster on one side of a surface than on the other, it is undergoing *velocity gradient flow* (*Mackin*, 1947, pp. 26—28). In other words, velocity gradient flow occurs when the velocity of a fluid is different along different stream lines in the fluid at any given time. Velocity gradient flow is shown with reference to both the *d* and *e* axes of a parallel stream tube in diagram *k*.

The nine stream tubes *a* through *i* drawn in the ring represent six different types of irrotational flow because, with 90° rotations about *f*, stream tubes *b*, *d*, and *g* equal *c*, *f*, and *h*, respectively. By adding velocity gradient flow with reference to the *e* axis to the nine stream tubes in the ring, *df* becomes the principal stream surface, and all nine tubes represent different types of irrotational velocity gradient flow. If velocity gradient flow with reference to *d* but not *e* were

Table 1. *Symmetry of irrotational flow types**

	Spherical $(\infty/m\,\infty/m\,\infty/m)$	Axial $(\infty\,mm)$	Ortho-rhombic $(2\,mm)$	Monoclinic (m)	Triclinic (1)
	$d, e, f =$ axes of ∞-fold rotation; $de, df, ef = m$	$f =$ axis of ∞-fold rotation; $df, ef = m$	$f =$ axis of 2-fold rotation; $df, ef = m$	$ef = m$	
Convergent flow					
1		c^2, where $c_d = c_e$ (tube a)	c^2, where $c_d \neq c_e$		
2				c^2g	
3					c^2g^2
4				pcg	
5					pcg^2
6			cp or pc (tubes b, c)		
7				cpg	
8					cpg^2
Parallel flow					
9	p^2 (tube e)				
10				p^2g	
11					p^2g^2 (tube k)
Compound flow					
12				dcg	
13					dcg^2
14			cd or dc (tubes d, f)		
15				cdg	
16					cdg^2
Divergent flow					
17			dp or pd (tubes g, h)		
18				dpg	
19					dpg^2
20				pdg	
21					pdg^2
22		d^2, where $d_d = d_e$ (tube i)	d^2, where $d_d \neq d_e$		
23				d^2g	
24					d^2g^2

* Symmetry is described with reference to the internal flow axes d, e, and f, disregarding the direction of flow along f.

added to the nine stream tubes, *ef* would become the principal stream surface, but the stream tubes, with 90° rotations about *f*, would be the same as those with a velocity gradient across *e*; therefore, no additional flow types are so produced. If velocity gradient flow with reference to both *d* and *e* were added to the nine stream tubes, a surface containing *f* but intersecting both *d* and *e* would become the principal stream surface, but, again, no additional types of flow would be defined. Nevertheless, if the fluid were anisotropic like rock, it is possible that the mechanism of flow might differ enough with direction so that the principal stream surface would have subordinate but significant velocity gradient flow within it; nine additional types of flow result. For this book, therefore, *df* is considered the principal stream surface with *f* the stream line, and the types of irrotational flow total 24.

Abbreviations are used here for the various flow types, following a suggestion by *F. Allan Hills*. They are formed by combining the first letters of the words that describe the flow with reference first to the *d* axis and second to the *e* axis; tube *g*, for example, illustrates type dp (divergent, parallel). When flow has the same character relative to both *d* and *e*, the single first letter, squared, forms the abbreviation; tube *a* illustrates type c² (convergent, convergent). When a velocity gradient exists across the principal stream surface *df*, g is added to the abbreviation, and when velocity gradients exist both within and across *df*, g² is added; for example c²g and c²g², respectively. The 24 types of irrotational flow, grouped according to symmetry, are listed in this manner in Table 1.

Combining rotational flow about *d*, about *e*, and about both *d* and *e*, with each of the 24 flow types listed in Table 1 (e.g., c²rg, c²r$_e$g, c²r²g, respectively) would result in 72 types of rotational flow. However, because three of the irrotational flow types (c², p², d²) do not each have three different types of rotational flow but instead have only one (e.g., c²r = c²r$_e$ = c²r², with appropriate rotations of *d* and *e*), the different types of rotational flow total 66.

Correlation with Strain Features

The first step in correlating folds with flow types is to identify the principal stream surface *df* and stream line *f*. The principal stream surface is defined as the surface across which the principal velocity gradient exists. In slip folds, it corresponds to the slip surfaces across which relative displacement of adjacent particles occurs; ideally, therefore, the principal stream surface parallels the axial surfaces of slip folds. In flexural-slip folds, the principal stream surface parallels the slip surfaces across which net relative displacement of the compositional layering occurs; it parallels the trend of the layering and not the axial surfaces of the folds.

Stream lines are defined as lines tangent to the velocity vector at every point in a fluid. Where adjacent particles with the same velocity in a fluid describe straight, parallel stream lines (type p^2), the particles move together, and no relative displacement of the particles occurs. In convergent flow (types c^2, pc), geometrically the stream lines simply become closer together in the direction of flow, but physically the paths of adjacent incompressible particles must merge, some particles moving faster than others in order to get in the same line. Therefore, on the scale of the particles, the paths of relative displacement may be irregularly oriented, but the vector sum of these paths in a portion of the fluid considerably larger than the particles themselves would be approximately parallel with the stream lines. Where adjacent particles moving with the same velocity describe curved, parallel stream lines (types p^2r, p^2r_e, p^2r^2), the particles move the same distance in a given amount of time, but they are displaced relative to each other because their stream lines are different in shape. In this situation their paths of relative displacement parallel the stream lines. In velocity gradient flow with straight, parallel stream lines (types p^2g, p^2g^2), adjacent particles move different distances and thereby become displaced relative to one another, also parallel with the stream lines. However, in divergent flow (types d^2, pd), geometrically the stream lines become more widely spaced in the direction of flow, but physically the paths of incompressible particles must depart from those of neighboring particles, the net relative motion being *perpendicular* to the stream lines.

When these simple flow types are combined to produce the remaining types defined in the foregoing section, they are combined parallel with their stream lines (*f*). Therefore, in all combinations of parallel, convergent, rotational, and velocity gradient flow, the paths of relative displacement of adjacent particles parallel the stream lines. However, in all types with a component of divergent flow, the paths of relative displacement do not parallel the stream lines.

The slip lines that constitute the principal concern of Chapter III correspond to the paths of relative displacement that parallel the stream lines in velocity gradient flow, regardless of the character of the other components of flow. The slip lines deduced from flexural-slip folds correspond to the paths of relative displacement between adjacent beds or layers, and the slip lines deduced from slip folds correspond to that component of the paths of relative displacement between adjacent particles contributed by the velocity gradient(s).

Therefore, the methods described in Chapter III can be used on both kinematic fold types to find orientations of the stream lines (as slip lines) of the flow in which the folds developed.

The final step of determining the corresponding flow type is essentially a qualitative analysis of the strain relative to the principal axes of flow. The details of this step will differ for different associations of strain features, different geologic situations, etc.

The tundra landslide described in the preceding chapter serves as a simple illustration. It is assumed at the outset that the tundra sod was incompressible during flow. The sod-bedrock interface can be identified as the principal stream surface df because the sod slid downward along the bedrock surface, localizing a sharp velocity gradient across it; however, no evidence of a velocity gradient within the sod was recognized. The mean downslope direction can be identified as the mean stream line f. While moving downslope, the sod developed folds which, if averaged over the whole landslide, resulted in a thickening of the sod and a divergence of flow with respect to e. It seems safe to conclude that the sod shortened perpendicular to the hinge lines of the folds, approximately parallel to df because of the parallel geometry of the folds and their similarities to flexures in a rug. Furthermore, because the axial orientations of folds are distributed from perpendicular to nearly parallel to f within df, the sod shortened with respect to both d and f, and flow therefore converged with respect to d. This conclusion is compatible with the fact that the landslide moved within a convergence in slope, as indicated by attitudes of the tundra surface around the slide (Fig. 15a), and with the fact that the centers of the two separation angles defined by folds on opposite sides of the landslide are oriented a few degrees to opposite sides of the mean downslope slip line (Fig. 17b, c). The slope of the tundra surface within the landslide averages 4° lower than that around the slide, indicating that the sod may have rotated slightly during flow. This rotation would correspond to stream lines curved about d but not about e; no evidence of rotation about e was observed. It is concluded, therefore, that the strain features in the tundra landslide developed during compound, velocity gradient flow with or without minor rotation, namely cdg or cdrg.

V. Trollheimen's Rock Units

The rocks of Trollheimen are divided into three groups: granitic gneiss, quartzite, and interlayered amphibolite and mica schist. The granitic gneiss was named the Trollhetta granite by *Holtedahl* (1938, p. 50) for its occurrence in Trollhetta, a peak located less than a kilometer north of the area mapped. The quartzite sequence is called the Gjevilvatn group because of its good exposure in the southwestern part of the area along Gjevilvatnet and upon the prongs of Gjevilvasskammene. The mica schist and amphibolite section is called the Blåhø group, named for the mountain Blåhø, east and north of Gjevilvasskammene.

Subdivisions of the three groups are listed in the legend of the geologic map (Pl. 1), and the abbreviated lithologic descriptions given there largely suffice for the present structural analysis. However, one of the units within the Gjevilvatn group, the Indre Kam formation, carries the burden of this book by exhibiting in its more quartz-rich layers most of the folds considered here. For this reason it is described more fully in the following paragraphs.

The Indre Kam formation consists of two rock types, feldspathic quartzite and augen gneiss. The relative percentages of the two differ from place to place, but the augen gneiss seldom exceeds 20 percent. The gneiss is composed of white and pale pink microcline porphyroblasts in a dark grey matrix of biotite-plagioclase-microcline-quartz schist. The augen measure as much as 50 centimeters in length and 10 centimeters in height and width, though they range down to microscopic size and nearly equal dimensions. The equidimensional augen are large, subhedral single crystals, whereas the highly elongate augen are aggregates of small subhedral and anhedral crystals. Augen gneiss generally occurs throughout the unit in sharply defined layers, distinct from the body of feldspathic quartzite. The layers are not always found in the same positions within the formation, though their wide lateral persistence is characteristic; in most areas, for example, an augen gneiss layer is found at the unit contact with the Kam Tjern formation, but in some areas it is conspicuously absent.

The quartzite that makes up the other 80 percent or more of the Indre Kam formation is characterized by fine, even compositional layering and parallel schistosity which allow the rock in weathering and mass wasting to break into large, thin sheets; this attribute has

Pl. 3. Flagstone (feldspathic quartzite) of the Indre Kam formation, Riaren, Trollheimen. Rock hammer indicates scale. *Top*. Massive and flaggy habits exhibited by the same rock type on opposite sides of joint surfaces. *Bottom*. Disrupted flags

won it the name "flagstone" and has given it a commercial importance for construction in many parts of Norway (Pl. 3). A chemical
analysis of the flagstone is reported by *Barth* (1938, p. 64, Table 4),
and its mineralogic composition is given by the modal analyses in
Table 2. The rock is light grey and striped, being made up of milli-
meter-thick layers of quartz and feldspar separated by screens of
muscovite and a little biotite. It is fairly uniform in appearance
throughout the unit, though it becomes more gneissic as the Svarthammer formation is approached.

Table 2. *Modal analyses of flagstone, Indre Kam formation*

Constituents	Localities and counts					Average,
	Riaren	Blåhø, NW	Hemre Kam	Lang tjern	Nonshø	%
Quartz	533	515	432	489	524	50
Microcline	158	205	90	133	197	16
Plagioclase	149	154	106	80	106	12
Muscovite	100	65	311	164	77	14
Biotite	9	22	10	12	10	1
Epidote	34	25	36	112	67	6
Hematite	5	9	11	4	11	1
Sphene	4	4	0	6	8	Trace
Apatite	0	1	4	0	0	Trace
Chlorite	8	0	0	0	0	Trace
Total	1000	1000	1000	1000	1000	100

The Gjevilvatn group and parts of the Blåhø group have been
measured with hand level, compass, and tape on Gjevilvasskammene; unit thicknesses are summarized in Table 3. The section begins
where Rensbekken flows into Gjevilvatnet, and the first rock type is
a feldspathic and gneissic quartzite, intermediate between quartzite
and granitic gneiss. The section was measured up Rensbekken to the
mouth of the hanging valley, then carried eastward along strike to
Hemre Kam, and finally up and along Hemre Kam to Kam tjern,
where the structure becomes increasingly complex and the units begin
to be repeated by folding. The section ends in oligoclase-quartz schist
of the Rensbekk formation.

The actual thickness in such a measured section in high-grade
metamorphic rocks have little meaning when one considers the various mechanisms of tectonic thinning and thickening that may have
operated in its different parts during recrystallization, but the rel-

ative thicknesses permit the calculation of relative densities of the units. For this purpose, rock samples were collected within the formations to represent all the rock types encountered in the section, and their densities were determined dry (without allowing water to permeate the rock) by weighing each one, observing the volume of water each displaces, and dividing the weight by the volume; 44 samples were subjected to this treatment. Values for individual samples have been combined into values for the rock units by multiplying the density of a sample by the thickness of rock it represents in the measured section, summing the products for a whole unit, and dividing by its thickness. The results, shown in Table 3, are obviously no more than indications of the present relative densities of the rock units.

Table 3. *Thicknesses and densities of rock units on Gjevilvasskammene*

Rock units		Thickness, meters	Density, g/cm³ (in parentheses, number of samples)			
Blåhø group	Rensbekk fm	200+	2.90	(7)	2.95	(16)
	Kam Tjern fm	324	3.00	(9)		
Gjevilvatn group	Indre Kam fm	340	2.75	(3)		
	Svarthammer fm	481	2.76	(14)	2.69	(26)
	Midtre Kam fm	12	2.61	(3)		
	Hemre Kam fm	927	2.63	(6)		
Trollhetta granite		—			2.72	(2)

In the preceding paragraphs, the granitic gneiss is referred to first, the quartzite second, and the schist-amphibolite sequence third, with the implicit assumption that the gneiss is oldest and the schist-amphibolite is youngest. One argument for this relationship is their configuration on the northern flanks of Storlifjellet, Mellomfjellet, and Gjeithetta, in the northwestern part of the area. There, the three units are found in an apparently simple homocline. The granitic gneiss has the lowest structural position, and the amphibolite-schist sequence the highest, indicating that the gneiss is older than the schist and amphibolite, as long as the premise — always tenuous in this tectonic setting — of simple structure and an uninverted sequence is correct. The same argument applies to their similar configuration on Gjevilvasskammene where the quartzite on the shore of Gjevilvatnet is gneissic and probably gradational into underlying granitic gneiss, and the quartzite in the higher reaches of Gjevilvasskammene is overlain by mica schist and amphibolite.

A second and more convincing argument is the analogy to rocks of the Trondheim basin. Fossils have not been found in Trollheimen, nor have they been found in any of the other high-grade rocks that lie to the north, south, or west and are traceable into the area. To the east, across the border fault, however, unmetamorphosed and low-grade rocks of the Trondheim basin have yielded fossils from several levels. There, the stratigraphic column consists of a granitic gneiss complex, which in part is overlain by quartzite, overlain in turn by a complex of eugeosynclinal sediments containing volcanic lavas and tuffs (*Carstens*, 1960, pp. 4—10; *Strand*, 1961, p. 162). The eugeosynclinal complex has been divided into five groups, of which the upper three (Lower Hovin, Upper Hovin, and Horg) range in age from Middle Ordovician to Silurian. The lower two groups ("Røros" and Støren) have yielded no fossils, but are considered Cambrian and Lower Ordovician by correlation with the fossiliferous Cambro-Silurian miogeosynclinal sedimentary rocks farther south in Norway. The underlying quartzite is supposed to be the metamorphosed equivalent of the feldspathic sandstone, sparagmite, which outcrops in the nappe region southeast of the core zone and which is Eocambrian in age. The granitic gneisses are presumably the Precambrian basement complex. On the basis of a short reconnaissance trip through Trollheimen, *Holtedahl* (1938) suggested that the schist-amphibolite sequence is correlative with the Cambro-Silurian Trondheim schists and the quartzite sequence with the Eocambrian sparagmite; the nature of the Trollhetta granite, being either a part of the Precambrian basement or a Caledonide intrusive, was uncertain to him. His correlation is accepted in a general way, today. The Trollhetta granite probably fits into the Precambrian (?) basement complex of Holmsen (*Strand* and *Holmsen*, 1960, p. 9); at least part of the quartzite sequence (Gjevilvatn group) is equivalent to the Eocambrian sparagmite; and the interlayered amphibolite and mica schist (Blåhø group) is correlative with the lowermost groups of the Cambro-Silurian eugeosynclinal rocks of the Trondheim basin.

VI. The Sahlfold Facies

Three cross sections that radiate from Storlifjellet in the western part of the area covered by Plate 1 are shown in Fig. 26; the Indre Kam formation, consisting primarily of flagstone, is drawn in black. The sections show the general forms and relative locations of the major structural complexes in southeastern Trollheimen: Svarthammer anticline, the eastern homocline, and Riar basin. Structures in the rocks east of Svarthammer anticline are distinct for the most part from structures west of the anticline, and therefore they are described in separate chapters of this book. The present chapter concentrates on structures of the sahlfold facies, which developed during an early and major stage in the formation of Riar basin.

Macroscopic Folds in Riar Basin

A large part of the basin is underlain by mica schist and amphibolite of the Kam Tjern formation, intricately interfolded on all scales with flagstone and augen gneiss of the Indre Kam formation. Though the map pattern described by compositional variations is intricate, the relationship exhibited by axes of macroscopic folds throughout the basin is simple. These axes (β) have been determined by plotting together on an equal-area net attitudes of foliation from different outcrops within individual areas of about 5000 square meters each. The folds are cylindroidal, and their axes vary regularly in direction and value of plunge from one locality to the next. The axes describe a conical pattern with a vertical central axis in the vicinity of the lake, Fosdals tjern. At Fosdals tjern fold axes are vertical, whereas at greater and greater distances from there they plunge toward the lake at smaller and smaller angles until the crest of Svarthammer anticline is reached; there, the shallow axes plunge in a direction nearly perpendicular to the axis of the anticline. Furthermore, the amount of contortion displayed by the foliation decreases outward from the center; near the lake foliation is strongly folded, whereas near the crest of the anticline foliation is relatively planar. The whole basin within the arcuate anticline is a cone of converging folds that become steeper and more closely spaced to a vertical central axis at Fosdals tjern.

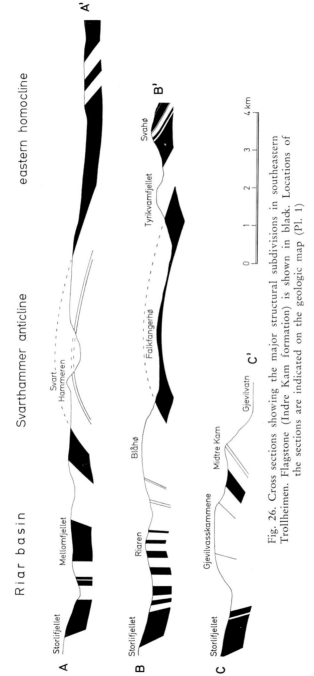

Fig. 26. Cross sections showing the major structural subdivisions in southeastern Trollheimen. Flagstone (Indre Kam formation) is shown in black. Locations of the sections are indicated on the geologic map (Pl. 1)

One of the macroscopic folds outlined on the map by the flag-stone-amphibolite contact bears striking resemblance to a hand in map pattern (Pl. 1). Fosdals tjern conceals its "palm," but the four "fingers" are exposed on Riaren, the bifurcated "thumb" on Blåhø, and the narrow "wrist" on Mellomfjellet. The "palm" contains the cone axis defined by axes of macroscopic folds, and fold axes at the extremities of each "digit" and on the "wrist" plunge toward this cone axis at angles of 70° and more. Following the convention of viewing folds in profile, the map itself is the best approximation to a profile section of this conical structure because the cone axis is vertical.

Mesoscopic Structures in Riar Basin
Mineral Lineation

Foliation throughout the area contains a prominent mineral linea-tion which is best displayed by biotite, muscovite, specular hematite, and amphibole. The lineation varies continuously from one place to another and everywhere parallels (within 10°) the macroscopic fold axes defined by intersections of foliation in spherical projection. Consequently, the lineation has the same conical distribution as that displayed by the macroscopic fold axes. To portray this structure, the angle of plunge of the mineral lineation has been contoured at 10° intervals on a topographic base of the whole area showing geologic contacts (Fig. 27). Various patterns have been added to areas of steeper plunge, the darkest where the plunge is 80° and greater. Adequate control for the isogons is lacking in certain areas of the map, but this drawback is less serious for the area of Riar basin than for others. The isogons do not take the direction of plunge into account, though they are roughly perpendicular to it in the basin. The area enclosed by the 80° isogon, which represents the deepest part of the basin, contains Fosdals tjern as well as the "palm," the first three "fingers," and the base of the oversized fourth "finger" of the flagstone "handfold," and the lower value isogons surround the 80° isogon concentrically until the area of Svarthammer anticline is reached.

Southwest of Fosdals tjern, the 80° and 70° isogons are sharp-ly indented, and, to the northeast, all the isogons are slightly indented. These indentations indicate the presence of a gentle "culmination," which trends northeastward and intersects the cone axis at Fosdals tjern, and within which macroscopic fold axes and mineral lineations plunge at shallower angles than to either side. Apparently, however, the culmination has a negligible effect on the configuration of the rocks units; it may have changed the length of the fourth "finger" relative to the first three of the "handfold," and it may have added sharpness to the westward trend of Svarthammer anticline at the head of Slettå valley.

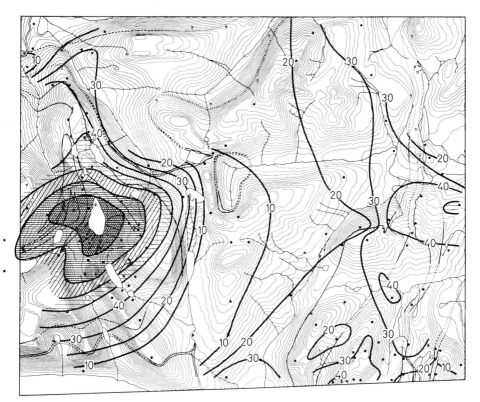

Fig. 27. Isogonic map of the plunge of the prominent mineral lineation, southeastern Trollheimen. Contour interval is 10°. Patterns locate areas of steeper plunge, the darkest where the plunge is 80° or more

In the whole western area, the mineral lineation plunges toward the environs of Fosdals tjern where the lineation is vertical, except on the northern flanks of Mellomfjellet and Gjeithetta and the southern slopes of Gjevilvasskammene. To the north, the direction of plunge of the mineral lineation gradually changes until it is westward in the valley of Folla north of Mellomfjellet. To the south, the lineation becomes difficult to observe on the southern prongs of Gjevilvasskammene, and, where it is measurable again near the shore of Gjevilvatnet, its direction of plunge has changed nearly 90° to barely

5*

north of east. As can be seen from the geologic map (Pl. 1), this east-west trend is a regional one that is found throughout Troll-heimen, exclusive of Riar basin.

Elongate Augen

Microcline augen in the Indre Kam formation display a spectacular lineation. In any exposure of augen gneiss, augen that are practically equant in dimensions can be seen beside augen with ratios of long to short axes greater than 10 to 1, and there are all gradations of elongation between these extremes. Long axes of the elongate augen are mutually parallel in any given exposure. Furthermore, they are rigidly parallel with the biotite lineation in the matrix. This biotite lineation is part of the mineral lineation described in the preceding paragraphs, and therefore the orientation and spatial relationships of the long axes of augen in Riar basin are the same as those described for the mineral lineation.

Boudinage

Boudinage is not seen in flagstone, gneiss, or quartzite of the Gjevilvatn group; rather, foliation in this uniform sequence of rock is characterized by extreme regularity. In contrast to the Gjevilvatn group, the interlayered amphibolite and mica schist of the Blåhø group exhibit a foliation so contorted by boudinage that the rock appears angular and grotesque in outcrop. In any small area it is possible to find boudin lines (or boudin axes) in practically every orientation, and the significance of a few scattered measurements is doubtful. However, any group of 50 boudins in the basin yields a strong preferred orientation of boudin lines at high angles to the mineral lineation and axes of macroscopic folds; the modes of such groups commonly occur at 85° to 90° to the lineation (Fig. 28a) but have been observed as low as 65° to 70° (Fig. 28b). A systematic investigation of this angle with location in the basin may yield some useful relationships but has not yet been attempted.

Sahlfolds

Sahlfolds exposed in flagstone within the basin are described in Table 4, of which certain aspects are illustrated in Figs. 29 to 33 and Plates 4 to 6. They can be characterized broadly as harmonic, similar folds, almost isoclinal on the average, with long, straight hinge lines and an uncommonly high degree of cylindricity. They exhibit a well-developed schistosity parallel to their axial surfaces and a prominent

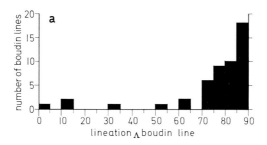

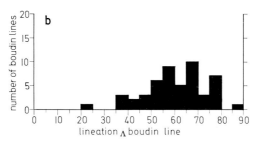

Fig. 28. Histograms showing the angular orientation of 50 boudin lines with respect to the local mineral lineation within mica schist and amphibolite in each of two areas in the valley of Kam tjern; *b* is taken 1 km. northwest of *a*

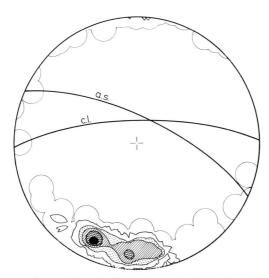

Fig. 29. Cleavage poles to 100 mica flakes from the limb of a sahlfold within the flagstone "thumb" on Blåhø. Compositional layering is labelled *c.l.*, and the axial surface *a.s.* Contours: 1, 5, 9, 13, 17, 19% per 1% area; maximum: 21%

Table 4. *Properties of sahlfolds in quartz schist (flagstone), Indre Kam formation*

Properties of an individual

---

1. Type of geometry	Similar[a] (Pls. 4, 5)
2. Nature of hinge and limbs	Small to moderate radii of curvature, broadly curved limbs (Pls. 4, 5)
3. Ratio of height to width	Range from 0.1 to 8.0 (cf. Fig. 4), based on $n = 65$
4. Ratio of depth to width	Range from 4.4 to 24.3 (cf. Fig. 6), based on $n = 22$[b]
5. Length and character of hinge line	Long and straight[c]
6. Cylindricity	Cylindrical[c] and cylindroidal; some conical (cf. Fig. 7)
7. Relation to cleavage	Foliation is folded; well-developed schistosity parallels axial surfaces[c] (Fig. 29)
8. Relation to mineral lineation	Fold axes parallel lineation within foliation in hinge areas; up to 15° from parallel to lineation within axial-surface schistosity[c] (Fig. 30)

Properties of a group

---

9. Mean and standard deviation of height-width ratios	Means of separate groups = 0.8 of $n = 14$, 0.9 of $n = 20$ (Fig. 31 a, b); composite mean = 1.2 of $n = 65$ (Fig. 31 c)[c]. Standard deviations = 0.57, 0.59 respectively; composite standard deviation = 1.16. (Finely foliated amphibolite: mean = 2.0 of $n = 8$; mica schist: mean = 1.4 of $n = 6$.)
10. Mean of depth-width ratios	Mean = 13.0 of $n = 14$ (Fig. 32a); composite mean = 14.1 of $n = 22$[b] (Fig. 32b)
11. Preferred orientation of fold axes	Linear distribution, with a slight spread in the plane parallel to their axial surfaces (Fig. 33)
12. Asymmetry	Normal (Pl. 5, top); with very large separation angles (150°+, Fig. 33b); some mixed (Pl. 6), with no separation angle defined (Fig. 33c)

---

[a] After the method of RAMSAY (1962a, pp. 310—311), thicknesses of eighteen individual layers have been measured parallel with the axial surfaces around eight sahlfolds in flagstone and mica schist in different parts of Riar basin. Most of the eighteen layers become thicker gradually from one limb of a fold to the other limb, with no indication of thinning at the hinge; some of the layers are practically the same thickness on both limbs as at the hinge; and only one layer of the eighteen is slightly thinner at the hinge than on the limbs.

[b] This number is small because the thought of taking such measurements occurred near the end of the last field season, not because the measurements are difficult to take.

[c] Diagnostic property, useful in distinguishing between different types of folds in Trollheimen.

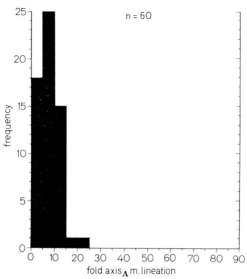

Fig. 30. Histogram of angles between sahlfold axes and the mineral lineation within sahlfold schistosity in Riar basin

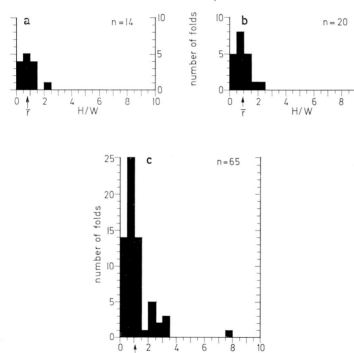

Fig. 31. Histograms showing height-width ratios of sahlfolds in flagstone. The mean values are indicated by arrows labelled $\bar{r}$. *a* Measured in a single outcrop within the major fold on Riaren, northwest of the "handfold". *b* Measured in a single outcrop within the "thumb" on Blåhø. *c* Composite histogram of all the ratios measured in flagstone in Riar basin

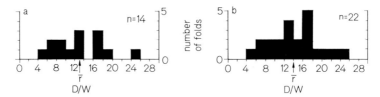

Fig. 32. Histograms showing depth-width ratios of sahlfolds in flagstone. The mean values are indicated by arrows labelled $\bar{r}$. *a* Measured in a single outcrop within the "thumb" on Blåhø. *b* Composite histogram of all the ratios measured in flagstone in Riar basin

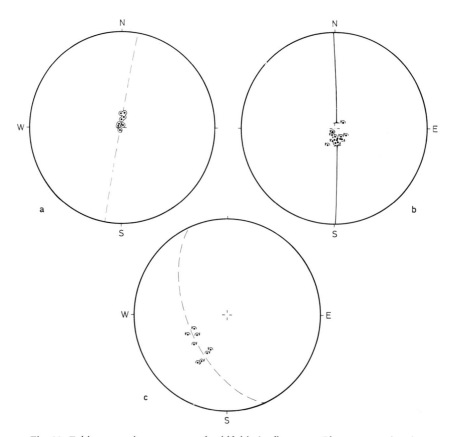

Fig. 33. Fold axes and asymmetry of sahlfolds in flagstone. Planes approximating the spread of axes are drawn as great circles. *a* 8 axes in the third "finger" on Riaren. *b* 15 axes in the major fold northwest of the "handfold" on Riaren; 161° separation angle. *c* 8 axes in the "wrist" on Mellomfjellet

Pl. 4. Sahlfolds in flagstone. *Top.* Profile view of a coupled sahlfold on Riaren. Rock hammer indicates size. *Bottom.* Large sahlfold on Riaren. Size given by sledge hammer

Pl. 5. Sahlfolds in Kamtjerndalen. *Top.* Two orders of folds showing normal asymmetry relations, in augen gneiss. *Bottom.* Folds illustrating typical similar geometry

Pl. 6. Sahlfolds showing mixed asymmetry relations resulting in fan forms, north-eastern Mellomfjellet. *Top.* Core of a fanfold in flagstone. *Bottom.* Large fan in interlayered flagstone and amphibolite

mineral lineation parallel to their fold axes. Their fold axes show a linear preferred orientation. Some of the commonly used names for sahlfolds are "similar folds," "reclined folds," "slip folds," and "flow folds," though these names are also applied to folds that are not sahlfolds.

Sahlfolds display the same distribution and orientation as the macroscopic folds and mineral lineation in Riar basin. They appear most common in the center of the basin and less and less common away from the center. Their axes plunge toward the vicinity of Fosdals tjern, those near the lake being vertical and those farther away from it shallower. Isogons of their plunges would show the same general form as the lineation-plunge isogons (Fig. 27).

Tubular Folds: Interference Structures

In nine localities within Riar basin, sahlfolds have been found associated with folds shaped like tubes, whose nearly parallel walls of foliation turn through 360° and close on themselves. These localities all lie well inside the basin, away from the surrounding Svarthammer anticline, though they do not appear to be confined to a particular, small area within the basin (Pl. 1). Five of the localities are in flagstone of the Gjevilvatn group, two in amphibolite of the Blåhø group, and two in schist of the Blåhø group. These structures are interpreted to be interference domes and basins formed by the superposition of sahlfolds on preexisting folds. They contain a wealth of structural information in their geometry that enables us to get some quantitative understanding of the flow environment in which sahlfolds developed. Therefore, three of the best exposed and most extensive areas of tubular folds, all in the "thumb" of the flagstone "handfold," were studied in detail. They are discussed in the following pages.

Area A. The eastern tip of the bifurcated "thumb" of flagstone here called area A, has been mapped with a plane table and alidade at scales of 1 to 100 and 1 to 20 (Pls. 7, 8, respectively [in pocket]; the larger scale map is a detail of the smaller). In walking across the part of the "thumb" that is represented in Plate 7, one would gain about 0.5 meter in elevation from east to west and almost 10 meters from north to south; the slope is gentle and even. The rock has only recently emerged from the melting glacier that still covers part of the "thumb" less than 100 meters to the north, and exposure is very good; the western half of the fold is close to 100 percent exposed, and the eastern half about 50 percent.

Outlining the fold in Plate 7 is the contact between flagstone of the Gjevilvatn group and amphibolite of the Blåhø group. The lines within the fold are individual layers of foliation that have been traced through the flagstone. This part of the "thumb" is synformal, the western limb being broader and more complex than the eastern one. The normal asymmetry pattern of folds of the next higher order is clockwise on the eastern limb and counterclockwise on the western limb. The map shows that those on the eastern limb are normal, but those on the western limb are both normal and reverse. Two relation-ships should be noted: (1) Traces of the axial surfaces of reverse folds on the western limb never cross the axial surface of this part of the "thumb" and enter the eastern limb. (2) All the tubular folds mapped in the area are found on the western limb, several of them associated with reverse, clockwise folds.

The tubular folds are cylindroidal and conical synforms and anti-forms, or basins and domes, connected by saddles. The general form of one of the basins (the large one in Pl. 8) is reconstructed in Fig. 34, and several of the folds and saddles are illustrated in Plates 9 to 11.

Prominent axial surfaces of sahlfolds have been traced through the tubular folds shown on the maps. When the forms of the tubular folds are examined with respect to these axial surfaces, it becomes apparent that the tubes are arranged in a rigid pattern. (1) The seven tubes in Plate 8 can be divided into four groups along their approxi-mately north-south trending axial surfaces. To the north are three domes (and possibly a fourth in the partly covered northeastern cor-ner) that define an east-west line, which is oriented at high angles to the axial surfaces. Southward from these three are two basins aligned in the same manner. Farther southward is a dome, and at the southern extremity is a single basin. (2) Within the northern group of domes, a saddle separates each dome from the next dome in line. Similarly, a saddle is found between the two basins to the south of the domes, and a saddle occurs beside both the dome and the basin farther to the south. (3) An axial surface traced through any basin in the map area either passes through a saddle in the next group of domes to the north or south or dies out before getting that far. In the same manner, an axial surface traced through any dome in the area either passes through a saddle in the next group of basins or it too dies out before getting that far. Therefore, the domes, basins, and saddles in Plate 8 occur at points of intersection in a two-dimensional, more or less rectangular (but somewhat distorted) grid. The dome, basin, and two saddles within the part of the fold marked *A* in Plate 7 are similarly arranged. This pattern is characteristic of two interfering sets of folds *(O'Driscoll,* 1962, 1964; *Ramsay,* 1962b, p. 468, Fig. 2).

One might conceive of such tubular folds forming as diapirs (as was assumed before the plane-table maps were made), but in that eventuality one might expect to find that the tubular folds were

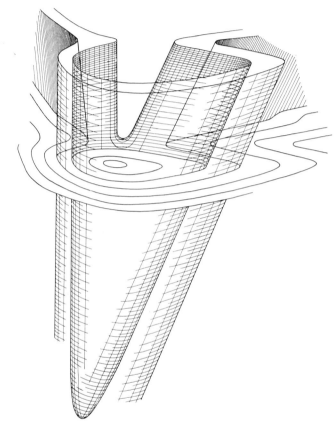

Fig. 34. Conical basin in flagstone, reconstructed in three dimensions from information in Plate 8

primarily domes, arranged in not so regular a pattern. Or, if they were primarily basins also arranged somewhat less regularly, they might conceivably be inverse diapirs. Nevertheless, domes appear to be no more important than basins, or vice versa, the observed pattern is clear, and the correspondence of this pattern with that expected of interfering fold sets is accepted here as conclusive evidence that the tubular folds are indeed interference structures.

Pl. 9. Views of a sahlfold interference basin in the flagstone "thumb" on Blåhø. This is the large basin shown in Pl. 8 and reconstructed in Fig. 34

Pl. 10. Two sahlfold interference domes and saddles in the flagstone "thumb" on Blåhø (Pl. 8).

Pl. 11. Two sahlfold interference saddles in the flagstone "thumb" on Blåhø
(Pl. 8)

6 Hansen, Strain Facies

The progressive change in the profile form of sahlfolds as they approach the tubular folds is compatible with this conclusion. The height-width ratio of any given sahlfold in profile increases along its axial surface from zero to a maximum value at a tubular fold, reverses its asymmetry across the tube, and, starting there with a maximum value, decreases to zero away from the tube in the opposite direction. The sahlfold that passes through the large basin that dominates much of Plate 8 well illustrates this point. (Its southern extremity can be found in Plate 7.) The attitude of the hinge line changes gradually from closely parallel with the steep mineral lineation near one end of the axial surface of the sahlfold, to horizontal within the tube and saddle, to steep in the opposite direction and parallel with the mineral lineation at the opposite end of the axial surface of the sahlfold; the fold axis varies practically $180°$ along the axial surface of such a sahlfold. These gradual changes in profile form and attitude of the hinge line occur with no indication of deformation or rotation, which might otherwise indicate that sahlfolds with average profile forms and parallel fold axes along their axial surfaces had been refolded into this configuration. Moreover, their axial surfaces, including the portions in the tubes, are planar or broadly curved, showing little sign of subsequent deformation. Rather, the whole configuration is suggestive of the strain associated with an average sahlfold imposed upon previously folded compositional layering, instead of being imposed upon the more common, relatively planar layering. Therefore, not only may we conclude that the tubular folds and saddles have resulted from the interference of two sets of folds, but we may also identify the later set as sahlfolds.

In order to analyze the geometry of these interference structures and deduce the orientations of the slip lines along which the sahlfolds developed, using the methods discussed in Chapter III, it is necessary to assume that the sahlfolds in this area were produced by slip. This assumption is made here with the following observations as support. Sahlfolds in the flagstone of Riar basin, including those in the flagstone "thumb," have the general form of folds usually assumed to be slip folds (similar; nearly isoclinal, average $H/W = 1$; harmonic, average $D/W = 14$; cf. Table 4). The paths of early fold axes rotated by sahlfolds are planar (as discussed in the following pages; Figs. 36a, 37d, 39a) and within $10°$ of parallelism with the average orientation of sahlfold axes in the area and with the axis of the macroscopic sahlfold, the "tip" of the flagstone "thumb" (Pl. 7). Theoretically, this geometry would result from slip folding and not from flexural slip (pp. 27–28; *Weiss*, 1955, 1959a; *Ramsay*, 1960). Furthermore, it seems impossible for such highly attenuated interference structures, like the one drawn in Fig. 34, to result from the flexing of previously folded layers across their early hinges, as any simple experiment with folded pieces of paper will demonstrate; rather, slip parallel with the late axial surfaces must accomplish most if not all of the refolding. Finally in support of the assumption that the sahlfolds in this area were produced by slip rather than flexural slip is the compatibility of the data presented in the remainder of this section.

The sahlfold axial surfaces that transect the largest basin and its adjacent saddle in Plate 8 attain their greatest separation near their intersection with the imaginary grid line connecting the basin center and the saddle; the same relationship is apparent for the basin and saddle along the same line to the east in Plate 8. In addition, three out of the four axial surfaces die out before they reach the east-west line connecting the domes and saddles to the north. These relationships, apparent in other portions of the map as well, indicate that the east-west grid lines, which represent axial surfaces of the early folds, had an effect on the sahlfolds superimposed upon them. An obvious effect is one of strength, the early hinges acting somewhat like struts relative to their limbs during refolding.

Two possible explanations of this effect come to mind: (1) The layering itself had additional strength parallel with the hinges because of its configuration, like the added strength of corrugated metals or cardboard parallel with the corrugations. If this explanation is applicable here, it indicates that flexural-slip folding was active during at least part of the development of the sahlfolds; pure slip folding occurs only when the layers have essentially no strength and behave passively. (2) The volume of rock containing the early hinges had more strength than the surrounding rock because the hinge areas localized the most intense deformation during early folding, and the rock became more compressed there than elsewhere. Consequently, any compression perpendicular to the axial surfaces of the sahlfolds during or subsequent to their development might be unable to compress the rock containing the early fold hinges as much as that containing the limbs, thereby producing wider spacing of sahlfold axial surfaces at early fold hinges than on their limbs. Unfortunately, the geometry was not investigated in enough detail to discover which, if either, of these two explanations is correct. Nevertheless, both are compatible with the conclusion that the final and main stage of development of the sahlfolds was accomplished by slip.

Eight attitudes of the schistosity that parallels the axial surfaces of sahlfolds in area A are shown in equal-area projection in Fig. 35a. These are all the complete attitudes obtained in the area, though

6*

numerous trends are shown on the two maps. They vary about 40°
in orientation and share a common axis near the vertical, which is
apparent in diagram *b* as the strong point maximum of the contoured
points of intersection of the eight planes in *a*. In diagram *c*, the 24

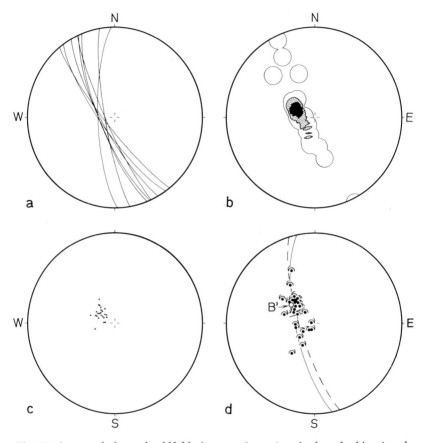

Fig. 35. Structural data of sahlfolds in area A. *a* 8 attitudes of schistosity. *b*
Contour diagram of 28 mutual intersections of the schistosity attitudes in diagram
a. Contours: 4, 18, 32⁰/o per 1⁰/o area; maximum: 46⁰/o. *c* 24 attitudes of the mineral
lineation; maximum: 88⁰/o per 1⁰/o area. *d* 29 fold axes; maximum: 52⁰/o per 1⁰/o
area. Separation angle: 4°

attitudes of the mineral lineation measured in the area display a
strong point maximum, located approximately where the maximum
in *b* is located. Therefore, the zone axis of the schistosity closely
parallels the mineral lineation.

All the axes of sahlfolds whose asymmetry is unequivocal and which are found on the western limb of this portion of the flagstone "thumb" are shown in Fig. 35d. The two axes connected by the solid but incomplete great-circle arc were measured along a single sahlfold axial surface on opposite sides of a tubular fold; the great-circle arc connects them through the horizontal to indicate the continuum of axial orientations through the tubular fold. The 29 fold axes are grouped into a point maximum, the center of which plunges about 65° to the northwest, similar in direction to the zone axes of the schistosity and the mineral lineation but shallower in plunge. Fold axes are shown at two places along the axial surface of the macroscopic fold in Plate 7; they are oriented within 1° of each other, and their average orientation, labeled B' in diagram d, is near the center of the point maximum of the mesoscopic folds. The point maximum spreads toward a great-circle girdle (approximated by the dashed line), which lies within the spread of orientations of the schistosity (diagram a). The axes of folds of opposite asymmetry occur in different parts of the girdle; the counterclockwise ones (normal in asymmetry to the macroscopic fold) plunge to the northwest and the clockwise ones (reverse) plunge to the west and southwest. The two groups overlap by 4°, thereby defining a separation angle of that size. The separation angle approximately parallels the zone axis of the schistosity and the attitudes of the mineral lineation (diagrams b, c, respectively).

Two tubular folds are well enough exposed in this area to permit determinations of early fold axes whose present orientations are due to rotation during subsequent sahlfolding. One of these is the large basin in Plate 8. Attitudes of the foliation were measured along the sides of the basin to determine the axes of curvature where the east-west grid line intersects them. Theoretically, these axes of curvature represent the fold axes (β) of the early synform; they are plotted as squares in the northwestern quadrant of Fig. 36a. The attitude of the horizontal grid line connecting the basins and saddles was taken as a third fold axis of the early synform. All three fold axes lie along the same great circle. The same procedure was followed with the dome just north of the large basin. Three fold axes of the early antiform are plotted as triangles on the same diagram; they, too, define a great circle. It is interesting to note that both the early synformal and anti-formal hinge lines have been rotated nearly 180° toward parallelism on opposite limbs of the later sahlfolds. Intersections of the great circles with attitudes of the axial surfaces of their respective interfering sahlfolds are shown as open circles in Fig. 36a. In accordance with the discussion of slip folding of a preexisting, straight lineation (pp. 27–28; *Weiss*, 1955, 1959a; *Ramsay*, 1960), these intersections

represent solutions for the orientations of the slip lines by which
the sahlfolds developed.

Two rotation paths of the early hinge lines having been determined, it is now possible to return to the sahlfold axes in Fig. 35*d*
and evaluate their separation angle. The sahlfolds represented there

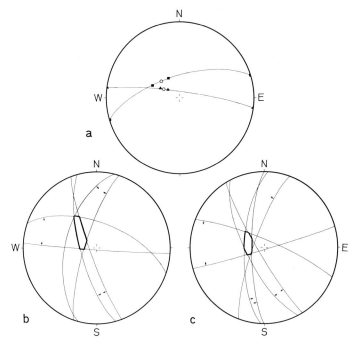

Fig. 36. Solutions for orientations of sahlfold slip lines in area A. *a* Squares
represent early fold axes (β) on the sides of the large basin in Plate 8; triangles
represent the same on the dome just north of the basin. Slip-line solutions are
shown as open circles. *b* and *c* Great circles represent attitudes of foliation around
the basin *(b)* and dome *(c)* of diagram *a*. Arrows indicate to which sides of the
foliation the stacking axis of each closure is oriented. They determine the polygon
(heavy lines) that includes the slip-line orientation for each closure

belong to a single order of folds that have developed on the limb of
a lower-order fold, the eastern "tip" of the flagstone "thumb." We
know from our earlier consideration of the geometry of superposed
slip folds of multiple orders that their separation angle may not
include the orientations of the sahlfold slip lines (pp. 47—50). Nevertheless, the maximum angular difference (Δ_{max}) between the slip-line
orientation and the separation angle can be determined with the
aid of the equation on p. 50. In order to do this, it is necessary

to know two other angular relationships, the angle θ between the sahlfold axial surfaces and the planar rotation paths of the early hinge lines, and the angle Φ between the sahlfold axial surfaces and the mean attitude of compositional layering on the western limb of the lower-order fold. Obtaining reliable values for these angles is itself a task of rather heroic proportions because, as we have seen from the maps and diagrams, the compositional layering is intensely folded, and both the sahlfold axial surfaces and the rotation paths of the early hinge lines display wide variations in attitude. Nevertheless, approximations that we hope are reliable have been obtained by using the average of thirteen attitudes of foliation where it is relatively planar, the average of eight attitudes of the schistosity that parallels the axial surfaces of sahlfolds (Fig. 35a), and the two rotation paths determined for the early hinge lines (Fig. 36a). The value obtained for Φ is 8°; the two values of θ are 87°, using the rotation path deduced from the line of basins, and 59°, using that from the domes. Choosing the value of θ that results in the higher value of Δ_{max}, we arrive at a value of 5° for Δ_{max}. Therefore, if the separation angle were enlarged at both ends by 5°, it would then include the slip-line orientation of the sahlfolds within its 14° spread.

Perhaps it should be made explicit at this point that the extended separation angle of 14° is an independent solution for the orientation of sahlfold slip lines. Although this solution depends upon the attitude of the planar rotation path of an early hinge line, just as the two solutions in Fig. 36a do, it does not depend upon the two solutions themselves, any more than they depend upon this one.

Attitudes of the foliation were measured around two tubular folds, the largest basin and the largest dome in Plate 8. In spherical projection, they show that the tubes are conical by defining somewhat regular polygons of intersections (outlined by heavy lines, Fig. 36b, c). Both these structures in profile (approximated by the plane of Plate 8) are convex outward in all directions and therefore fall into group I of Fig. 13. Thus, according to our earlier discussion of interference structures of this type (pp. 28–32), their polygons of intersections contain the orientation of the sahlfold slip lines.

In summary, therefore, five solutions for the slip-line orientations of sahlfolds have been obtained by three different methods in this area. A visual comparison of these solutions (Figs. 35d, 36) reveals that they plunge about 60° to 80° to the northwest. Furthermore, they are approximately parallel with other linear elements in the area — the zone axis of the schistosity, the mineral lineation, and the fold axes (Fig. 35).

Area B. Area B is located on the western edge of the flagstone "thumb," 425 meters northwest of area A. It is approximately one-fifth the size of that area and more poorly exposed. Much of the

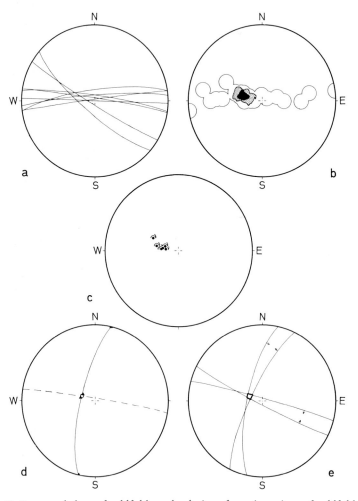

Fig. 37. Structural data of sahlfolds and solutions for orientations of sahlfold slip lines in area B. *a* 8 attitudes of schistosity. *b* Contour diagram of 28 mutual intersections of the schistosity attitudes in diagram *a*. Contours: 4, 18, 32% per 1% area; maximum: 46%. *c* 6 fold axes; maximum: 83% per 1% area. *d* Triangles represent early fold axes *(β)* on the sides of a basin. The slip-line solution is shown as an open circle. *e* Great circles represent attitudes of foliation around the basin of diagram *d*. Arrows indicate to which side of the foliation the stacking axis of the basin is oriented. They determine the polygon (heavy lines) that encloses the slip-line orientation

rock is broken and frost-heaved, and reliable orientation data were therefore difficult to obtain. The structural relationships observed in the area show striking similarity to those found in area A, and the arguments used in the discussion of those structures apply to the structures in this area as well.

Schistosity shows a similar spread in attitude and a steep common axis (Fig. 37a, b). No observations on the mineral lineation were recorded, but the six sahlfold axes shown in diagram c exhibit a strong linear preferred orientation that parallels the zone axis of the schistosity. All the folds are clockwise in asymmetry, reverse to the flagstone "thumb," and the separation angle they define is large and therefore not useful (161°, not drawn).

Two basins and a saddle along a single line were the only interference structures found in this area. One of the basins was well exposed and apparently in place, and, following the procedure used in area A, three axial orientations of the early synformal hinge were obtained (Fig. 37d). The intersection of the great circle that fits them (solid in the diagram) with the axial surface of the interfering sahlfold (dashed) is a solution for the orientation of the sahlfold slip lines. A second solution was obtained by measuring attitudes of the foliation around a tubular fold that is convex outward in all directions in profile (the well-exposed basin in this case), and then identifying the polygon in spherical projection that contains the slip-line orientation of the sahlfolds (heavy lines, diagram e).

It is apparent from comparison of the diagrams in Fig. 37 that the zone axis of the schistosity, the fold axes, and the two slip-line solutions are approximately parallel.

Area C. Area C is located on the western side of the flagstone "thumb," near its base, 625 meters northwest of area A and 240 meters west and slightly north of area B. In size and exposure, it is comparable to area A. As in area B, the observed structural relationships are essentially the same as those found in area A, and the arguments in the discussion of those structures apply to these as well. A wider variety of interference structures was found in this area; domes and basins belonging to both groups I and II by shape of their profile sections were seen (cf. Fig. 13). Furthermore, at least twice as many interference structures were found in this area as in the other two combined.

The linear elements measured in area C are mutually parallel (Fig. 38). All the recorded sahlfold axes are clockwise and reverse to the lower-order fold, as in area B, and their separation angle is similarly large (164°, not drawn).

Attitudes of early hinge lines rotated by subsequent sahlfolding were obtained from three lines of interference structures (Fig. 39*a*), one from an early antiformal hinge whose interference domes belong to group II (square symbols in the diagram), and one each from a

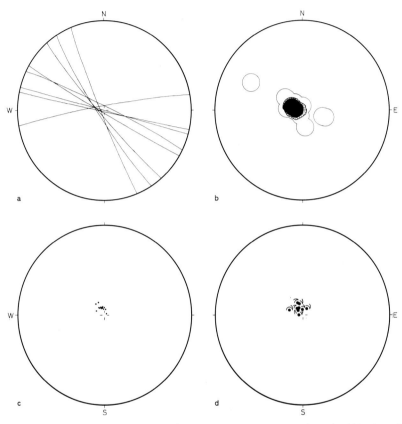

Fig. 38. Structural data of sahlfolds in area C. *a* 8 attitudes of schistosity. *b* Contour diagram of 28 mutual intersections of the schistosity attitudes in diagram *a*. Contours: 4, 18, 32% per 1% area; maximum: 75%. *c* 14 attitudes of the mineral lineation; maximum: 100% per 1% area. *d* 10 fold axes; maximum 100% per 1% area

synformal and an antiformal hinge whose interference tubes belong to group I. Although their rotation paths are far from parallel, their intersections with the axial surfaces of their respective interfering sahlfolds are only a few degrees apart. These intersections constitute solutions for the orientation of the sahlfold slip lines.

Structural elements obtained from a basin with triangular profile are shown in diagram *b* of Fig. 39. The axial surface of the interfering sahlfold is drawn as a dashed great circle, and an attitude of the foliation measured on the flat side of the triangle where it is crossed by the sahlfold axial surface is drawn solid. The intersection of these two planes (open circle) is taken as a fourth solution for the slip-line orientation of the sahlfolds, following the conclusions reached earlier in the discussion of the geometry of this type of interference structure (p. 30; Fig. 13*e*, *h*). Two additional solutions from the measured attitudes of foliation on the flanks of two domes of group I are shown in diagrams *c* and *d*.

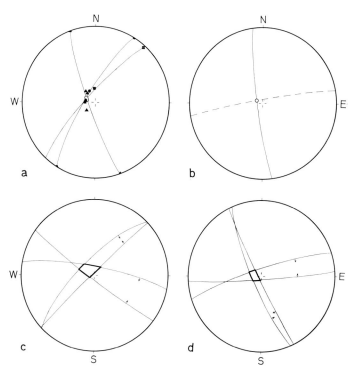

Fig. 39. Solutions for orientations of sahlfold slip lines in area C. *a* Early fold axes (*β*) on the sides of two domes and a basin are shown as dots, triangles, and squares, each type of symbol representing a different closure. Slip-line solutions are shown as open, coalescing circles. *b* Foliation (solid) and sahlfold axial surface (dashed) on a basin of triangular profile. Slip-line solution is shown as an open circle. *c* and *d* Great circles represent attitudes of foliation around two domes. Arrows indicate to which sides of the foliation the stacking axis of each dome is oriented. They determine the polygon (heavy lines) that encloses the slip-line orientation for each dome

As in areas A and B, the linear structural elements and the solutions for the orientation of the sahlfold slip lines are approximately parallel in area C (Figs. 38, 39).

Comparison of Data. The first three diagrams in Fig. 40 show the solutions for the slip-line orientations of sahlfolds in the three areas — five solutions from area A in diagram *a*, two from area B in *b*, and six from area C in *c*. Agreement among the solutions obtained by different methods within individual areas seems good, though some

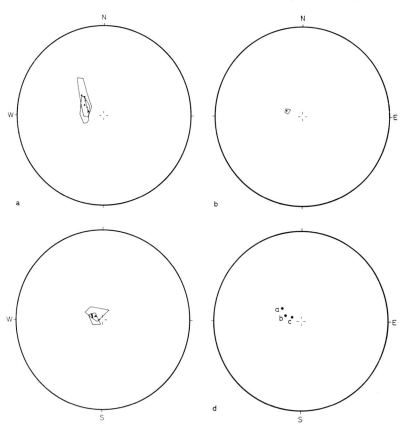

Fig. 40. Summary diagrams of slip-line solutions of sahlfolds in areas A, B, and C. *a* Five solutions from area A. Composite arc represents the separation angle from Fig. 35*d* and, added to both ends, the value of Δ_{max} calculated from the equation on page 50. Dots represent the solutions from Fig. 36*a*, and the polygons are taken from Fig. 36*b* and *c*. *b* Two solutions from area B, taken from Fig. 37*d* and *e*. *c* Six solutions from area C. Dots represent the three solutions from Fig. 39*a*, the triangle from Fig. 39*b*, and the polygons from Fig. 39*c* and *d*. *d* Means of slip-line solutions from diagram *a* for area A, from diagram *b* for area B, and from diagram *c* for area C

scatter is apparent. From these solutions, a mean slip-line orientation has been determined for each area by averaging the orientations of points, the center of the separation arc, and the center of gravity of the polygons within individual areas; the resulting mean orientations are shown in diagram *d*. This simple way of obtaining a mean orientation of the slip lines in a given area from diverse types of solutions is not completely satisfactory because it assumes that a solution from one method is as reliable as a solution from another, which may not be true, and it assumes that every orientation within a field of orientations (like a polygon in spherical projection) has the same probability of paralleling the slip-line orientation as does every other orientation within the field, which likewise may not be true. Nevertheless, this simple averaging is used here because no other, more accurate way of obtaining the mean orientations by assessing these factors is known at present. Moreover, the rather symmetrical disposition of solutions from one method relative to those from another within the diagram suggests that the effect of these two factors can only be slight.

The angle between the mean slip-line orientations of areas A and B is 7°, and the angle between those of areas B and C is the same. The scatter among certain individual solutions within the areas, however, is about the same size, which places in question the significance of this small angle between the means. In order to test whether the mean orientations are identical, the following statistic *(Watson and Irving*, 1957, pp. 290—293) has been calculated subject to the assumptions in the previous paragraph

$$\omega = \frac{2\,(\Sigma\,N_i - p)}{2\,(p-1)}\,\frac{(\Sigma\,R_i - R)}{(\Sigma\,N_i - \Sigma\,R_i)}$$

in which the sample from the *i*th of *p* areas contains N_i slip-line solutions and has a resultant of length R_i, and

$$R = [(\Sigma l_i)^2 + (\Sigma m_i)^2 + (\Sigma n_i)^2]^{1/2}$$

where the direction cosines of a slip-line solution, specified by the dip, *I*, and declination, *D*, are $l = \cos I \cos D$, $m = \cos I \sin D$, $n = \sin I$. Values of ω may be referred to the *F*-ratio tables with $2\,(p-1)$ and $2\,(\Sigma N_i - p)$ degrees of freedom. The sample value obtained for ω, 17.28, indicates that the differences between mean orientations for the three areas are significant at the 0.005 level. That is to say, on the basis of the distribution of the 13 slip-line solutions from the three areas, it is expected that if 1000 such solutions were obtained by the same methods, only 5 of them would be impossible to identify by orientation relative to the mean orientations from their respective areas. Therefore, we may conclude that the mean orientations and the angular differences between the means are reliable.

Histogram *a* in Fig. 41 shows the angles between the mean slip-line orientation for area A and the orientations of the 28 mutual intersections of schistosity from the same area (Fig. 35*b*). Histo-

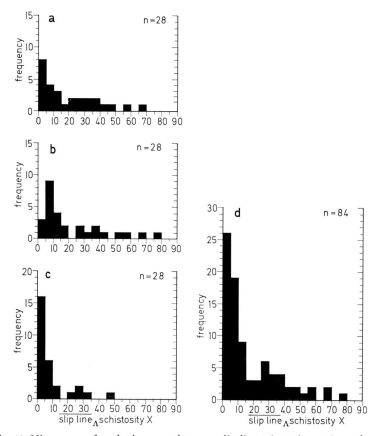

Fig. 41. Histograms of angles between the mean slip-line orientations (Fig. 40*d*) and the mutual intersections of sahlfold schistosity (Figs. 35*b*, 37*b* and 38*b*). *a* Area A. *b* Area B. *c* Area C. *d* Composite of *a*, *b*, and *c*

grams *b* and *c* show the same angles determined for areas B and C, respectively, and histogram *d* summarizes *a*, *b*, and *c*. The histograms in Fig. 42 show the angles between the mean slip-line orientations and attitudes of the mineral lineation in areas A and C, and the histograms in Fig. 43 show the angles between the mean orientations and the fold axes in all three areas. Modes for individual areas in the

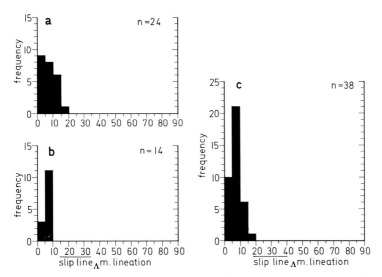

Fig. 42. Histograms of angles between the mean slip-line orientations (Fig. 40*d*) and attitudes of the mineral lineation (Figs. 35*c* and 38*c*). *a* Area A. *b* Area B. *c* Composite of *a* and *b*

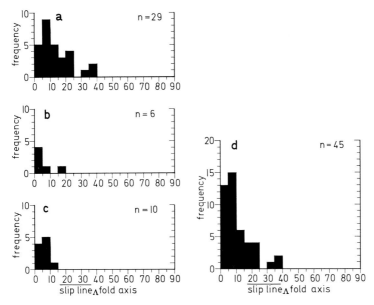

Fig. 43. Histograms of angles between the mean slip-line orientations (Fig. 40*d*) and sahlfold axes (Figs. 35*d*, 37*c*, and 38*d*). *a* Area A. *b* Area B. *c* Area C. *d* Composite of *a*, *b*, and *c*

three figures fall into the 0° to 5° and 5° to 10° class intervals. The mode in the composite histogram of slip-line — schistosity-intersection angles falls into the 0° to 5° interval (Fig. 41d), and those in the composite histograms of slip-line—lineation angles and slip-line—fold-axis angles fall into the 5° to 10° interval (Figs. 42c, 43d, respectively).

Although the histograms show spreads and modes of angles between mean slip-line orientations and the individual linear fabric elements, it is also of interest to determine the angles between mean orientations of the fabric elements and the mean slip-line orientation in each area. Mean orientations of the fabric elements have been calculated as axis 3 of the Dimroth ellipsoid (*Dimroth*, 1962a, 1962b, 1963), with use of the program VEC-4 of *Horace Winchell* and *William H. Scott*. The center of gravity of the most dense point maximum of spherically distributed data is determined by first finding the best-fitting great circle by a least-squares calculation and then finding its intersection with a second best-fitting great circle that contains the pole to the first. Mean slip-line orientations calculated in this manner for the three areas are identical to those calculated by the simple averaging procedure described previously; therefore the resulting mean orientations of fabric elements in these areas may be compared directly with the mean slip-line orientations. The results of this comparison are summarized in Table 5.

Table 5. *Angles between mean orientations (axis 3 of Dimroth ellipsoids) of slip-line solutions and of linear elements of sahlfolds in three areas of tubular folds*

a) Within areas	A	B	C
Schistosity X's ∧ min. lineation	4°	—	5°
Fold axes ∧ schistosity X's	7°(13°)[a]	2°	5°
Fold axes ∧ min. lineation	4°(10°)[a]	—	1°
Slip-line sols. ∧ schistosity X's	7°	6°	4°
Slip-line sols. ∧ min. lineation	5°	—	6°
Slip-line sols. ∧ fold axes	1°(6°)[a]	5°	5°

b) Between areas	A and B (425 meters apart)	A and C (625 meters apart)	B and C (240 meters apart)
Schistosity intersections	6°	10°	16°
Mineral lineation	—	11°	—
Fold axes	6°(13°)[a]	14°(18°)[a]	15°
Slip-line solutions	7°	13°	7°

[a] Angles in parentheses are based on the point maximum of eighteen fold axes in area A.

Implications for Sahlfolding. This study of three areas of tubular folds has revealed two important kinematic relationships of sahlfolds. One, summarized in the last line of Table 5a, is that the fold axes of sahlfolds make small angles (about 5°) with the slip-line orientations recorded in their geometry. Although this conclusion applies only to the sahlfolds in a relatively small part of Riar basin, it can be extended to much of the remaining part on the basis of less detailed observations on sahlfolds relative to tubular folds in the other widely separated areas on Mellomfjellet, in the valley of Kam tjern, and on Midtre Kam and Storlifjellet. In these areas, the fact that the axes of sahlfolds parallel the tubular folds indicates that the angle between the sahlfold axes and the slip-line orientation confined by the tubular folds must also be small. This relationship is somewhat surprising when considered in light of the commonly accepted belief that fold axes form approximately perpendicular — not parallel — to the slip lines along which they develop.

On the assumption that fold axes form perpendicular to slip lines, it is customary to consider the height of a fold in profile to be a good approximation of the distance that bedding has been displaced to form the fold. Since this assumption has proved false for sahlfolds, however, we might inquire what a better approximation of this displacement might be. From consideration of the geometry (Fig. 44), it can be shown that a value for the displacement of bedding during slip folding can be obtained from a displacement ratio

$$P = Y/H = \csc \alpha$$

where Y is the real displacement parallel with the slip lines, H is the apparent displacement or height of the fold, and α is the angle between the slip lines and the fold axis *(Hansen* and *Scott,* 1968). This treatment assumes the fold axis to be a fixed line of reference, but, if the fold axis were to rotate by differential slip within the slip surfaces during folding, the true displacement could differ somewhat from that calculated. Nevertheless, the simple displacement ratio, P, still gives a better approximation than the height of the fold. Therefore, taking $\alpha = 5°$, which is the average angle between the mean slip-line orientations and the means of sahlfold axes in the areas of tubular folds, the displacement ratio P for those folds is 11.5. Thus the displacement that actually occurred during the development of an average sahlfold in that portion of Riar basin is *an order of magnitude greater* than the displacement apparent in its profile section.

The other kinematic relationship revealed by this study is the downward convergence of the sahlfold slip lines. This convergence

can be seen in Fig. 45, where the mean slip-line orientations are
plotted on a skeletal map of the central part of Riar basin, showing
geologic contacts and lineation-plunge isogons. The convergence between areas A and B is 0.02°/meter and that between B and C is

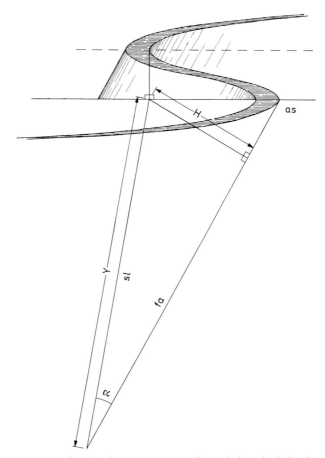

Fig. 44. Fold axis *(fa)*, slip-line orientation *(sl)*, and short-limb height *(H)* of a
hypothetical slip fold

0.03°/meter, which averages about 1°/40 meters along the line of
the three areas[7]. Cone axes of the tubular folds observed elsewhere
in the basin also converge downward, indicating that the sahlfold slip

---

[7] This value replaces that of 1°/50 meters reported earlier *(Hansen,* 1967b). The
angle of plunge of the slip lines was not taken into account in the earlier calculation.

lines, to which they are parallel within 15°, exhibit the same qualitative relationship throughout a rather large portion of the basin.

Tubular folds have been found to the north, east, south, and west[8] of the basin center (Pl. 1). In general, their direction of convergence is not only downward but also toward the area of steep plunges of the mineral lineation. Fig. 45 shows that the mean slip-line orienta-

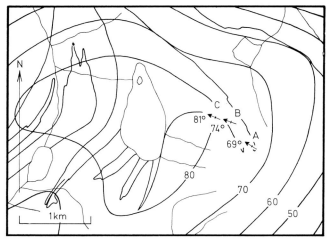

Fig. 45. Mean slip-line orientations for sahlfolds in areas A, B, and C, shown on a skeletal map of the central portion of Riar basin, including geologic contacts (Pl. 1) and lineation-plunge isogons (Fig. 27)

tions from areas A, B, and C to the east of the basin center plunge toward the island in Fosdals tjern, somewhat north of the center of the 80° isogon. Furthermore, cone axes of the tubular folds to the south of the basin center plunge slightly toward the west of the 80° isogon; presumably the sahlfold slip lines have a similar direction of plunge. It can be concluded, therefore, that sahlfold slip lines in much of Riar basin converge downward from all sides toward its steeper, central region, though slight deviations from this pattern exist.

Although we have seen evidence indicating that the sahlfold slip lines converge downward, and therefore diverge upward as well, we have yet to examine evidence that permits us to decide whether the rock moved upward or downward along those slip lines to form the sahlfolds. To answer this question, we turn to considerations of the macroscopic structure of the basin.

---

[8] The one to the west is located on southern Storlifjellet, just beyond the western border of Plate 1; its cone axis plunges toward the east at 65°.

The Form of Riar Basin

The usual designations of anticline and syncline are not readily applicable to macroscopic folds within the basin. Walking westward from the anticline at Svarthammeren (Pl. 1), one passes upward in the section through flagstone of the Indre Kam formation into schist and amphibolite of the Kam Tjern formation and then downward again into flagstone of the "handfold" at Fosdals tjern. Presumably the amphibolite-schist outcrops are localized by a syncline, but the flagstone at Fosdals tjern can be either anticlinal or synformal. The same ambiguity exists for the flagstone folds west of Fosdals tjern. Even their forms are unknown; though the portion of the "thumb" presented in Plate 8 is synformal, its plunge is quite steep, and the central parts of the "handfold" have vertical axes and can be either antiformal or synformal. Consequently, two alternative interpretations are examined in the next few pages, neither of which is proved by the known facts. The first assumes the flagstone folds to be anticlinal; the second assumes them to be synformal in a preexisting recumbent fold.

Anticline Hypothesis

Let us suppose that the flagstone folds are anticlines and that they connect beneath the mica schist and amphibolite with the flagstone in Svarthammer anticline. The three cross sections in Fig. 26 are reconstructed according to this hypothesis in Fig. 46. In addition, one possible reconstruction for the three-dimensional form of the "handfold" is drawn in Fig. 47. The "handfold" is the largest of the three anticlines and in three dimensions is suggestive of an irregular mushroom, overhanging on three sides but connected by a septum on the fourth side with the great flagstone mass of Storlifjellet. The other two anticlines are thin, bent plates, being a little broader in their upper portions than at their bases.

A case can be made for the equivalence of the "handfold" to a salt dome. The general shape of the anticline itself with its open, rim syncline to the northeast, east, and south is vaguely analogous. The conical distribution of mineral grain lineations and axes of macroscopic and mesoscopic folds, illustrated by the lineation-plunge isogons in Fig. 27, is typical of a salt dome that has begun "to mushroom" in the later stages of its ascent (*Trusheim*, 1957, Fig. 13). The mineral lineations and fold axes are vertical toward the center, like those in the Grand Saline salt dome *(Balk,* 1949, p. 1810; *Muehlberger,* 1959, p. 12). Comparison of Trollheimen's sahlfolds with folds in the Grand Saline and Jefferson Island domes *(Balk,* 1949, 1953; *Muehl-*

berger, 1959) reveals striking similarities in form. Of the eight properties of individual sahlfolds, only one is different for the folds in salt
domes: the salt folds are displayed by compositional layering and not by
foliation. Only their cylindricity is unknown; the other six properties
are the same: similar folding, small to moderate radii of curvature
in hinge areas and broadly curved limbs, variation from open to

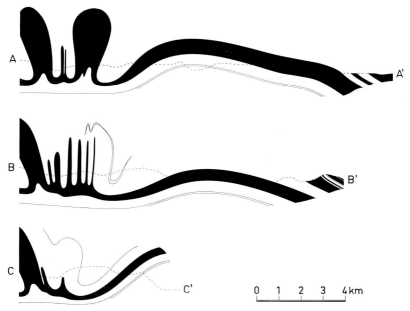

Fig. 46. Cross sections reconstructed according to the anticline hypothesis. Flagstone is shown in black. Location of the sections is indicated on the geologic map
(Pl. 1)

isoclinal (height-width ratios measured on fourteen of the clearest
coupled folds in *Balk's* 1949 Grand Saline map, Pl. 1, range from
0.4 to 2.0), harmonic (depth-width ratios from five of *Balk's* folds,
1949, Pl. 1, range from 5.8 to 12.5), extremely long and straight
hinge lines, and axes parallel to the mineral lineation. All four of the
group properties described for sahlfolds are correct for the folds in
salt domes: as a group, they border between open and isoclinal
(mean value of height-width ratios for the fourteen Grand Saline
folds is 1.0; standard deviation is 0.22), they are harmonic (mean
of depth-width values for the five folds is 9.9), they have parallel
axes, and most of them show a normal asymmetry relationship to
folds of the next lower order, but some show a reverse relationship.
Furthermore, the salt "closures" in the Jefferson Island dome *(Balk,*

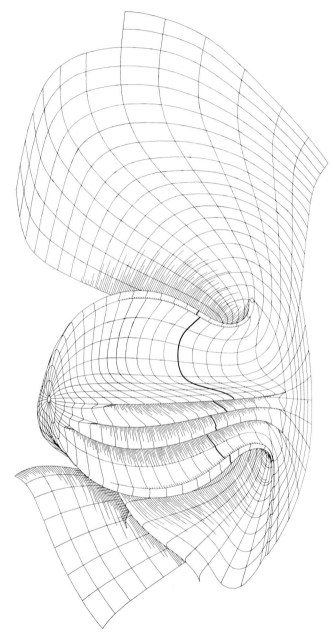

Fig. 47. Interface between flagstone and mica schist-amphibolite, reconstructed according to the anticline hypothesis for the form of the flagstone "handfold". The view is northward. Intersection of the land surface and the "handfold" is indicated by heavy lines. Down-dip lines roughly parallel mean orientations of the sahlfold mineral lineation and sahlfold axes. Cross lines are form lines, not contours

1953, Pl. 1, p. 2465, and Fig. 9) appear identical to the tubular folds mapped in Trollheimen (Pl. 8). The tubular folds are both antiformal and synformal, but the form of the salt closures is unknown. These identities of mesoscopic folds in Riar basin and in the two salt domes are the most compelling arguments for the "handfold"-salt dome analogy.

The density obtained for the flagstone sequence is 2.75 g/cm³ and for the schist-amphibolite sequence 3.00 g/cm³, at 1 atmosphere and about 20°C, dry (Table 3); their difference is 0.25, which is likely to be a minimum value for their difference under the conditions that obtained during flow. This is the same as the density difference between salt (2.1 to 2.2 g/cm³) and sediments (2.3 to 2.5 g/cm³; *Nettleton,* 1934, p. 1179, Fig. 1; *Goguel,* 1948, pp. 359 to 361; *Parker* and *McDowell,* 1955, p. 2400).

Calculations of salt-dome movement for the "handfold," using equations similar to those developed by *Goguel* (1948, pp. 360—362), involve too many unknowns to make them worth while. The most important unknowns are the shape and volume of the upward-moving mass. Was the irregularly shaped body of flagstone that makes up the "handfold" the active mass, or did all three flagstone anticlines with thin synclinal septa move upward together as a single, more regular unit, with the intricacies of the flagstone-mica schist contact inherited from a previous deformation?

The larger height-width ratios of sahlfolds in the schist and amphibolite, in comparison with those in flagstone (Table 4)[9], suggest that the relative viscosity of the schist-amphibolite sequence was lower than that of the flagstone during flow. Therefore, the relative viscosities of the domal material and the overburden in Trollheimen appear to have been the reverse of those of salt domes in sedimentary sequences. Movement of salt into a dome occurs only if the greater viscosity of the overlying materials is low enough to permit subsidence *(Nettleton,* 1934, p. 1184; *Dobrin,* 1941, p. 541). Movement of flagstone into the dome could occur only if the flagstone itself had a viscosity low enough to permit its upward flow; the amphibolite-schist sequence would then be capable of subsiding. However, because of the relative viscosities, perhaps one should expect a conical subsidence of the denser and more fluid schist and amphibolite and the consequent rising of the flagstone into a rim anticline. This, of course, is opposite to the form reconstructed here.

No indication of diapirism is seen in the rocks of the western area, yet it is characteristic of many salt domes. Perhaps the explana-

---

[9] This statement is based upon numerous qualitative observations in addition to the measurements reported in Table 4.

tion for the difference lies in the anomalous viscosity relationship. The viscosity of salt is so much less than that of the overlying sediments that the salt flows through the fracturing overburden; diapirism results. The viscosity of flagstone may have been only slightly greater than that of the overlying schist and amphibolite, and any flow initiated by the flagstone would have been readily complemented by flow of the amphibolite-schist sequence; conformable contacts would result.

Synform Hypothesis

Let us assume that the flagstone folds are synformal within a great recumbent fold. The complete separation by schist and amphibolite of flagstone within the basin from that of the surrounding homoclines and anticline may indicate that the two flagstone bodies do not connect beneath the basin but connect somewhere else, perhaps at a "root zone" beyond the western boundary of the mapped area. The synformal tip of the "thumb" (Pl. 7) may indicate that the whole "handfold" is synformal in a basin of schist and amphibolite; perhaps the associated folds to the west are also synformal. The continuous steepening of foliation and lineations from Svarthammer anticline to the center of the "handfold" without an apparent syncline in the Blåhø group may indicate that the syncline, if it does exist, is attenuated and isoclinal beneath an overlying recumbent anticline; it may also indicate the absence of a syncline, the upper flagstone body being repeated by a thrust fault. Early folds in the areas of tubes may represent nappe structures on which folds of a later event were superimposed.

More convincing evidence for the existence of a large recumbent fold in Trollheimen is obtained from adjacent areas. *Holtedahl,* who studied an area southwest of this one, concluded that the large-scale structure is a series of nappes, in places piled on top of each other, with basement-gneiss cores and flagstone-schist-amphibolite mantles (1950, p. 57). His cross section (Fig. 18, p. 56) includes part of the mountain Hornet, which lies directly south of Gjevilvatnet and which is gently synformal with a sequence, from bottom to top, of basal gneiss, flagstone, mica schist and amphibolite, and flagstone again. He tentatively interpreted the schist-amphibolite section to be a synclinal septum lying under a nappe of flagstone. The sequence on Hornet corresponds exactly to the sequence of rocks north of Gjevilvatnet on Blåhø and Gjevilvasskammene and suggests that the uppermost flagstone layers are indeed parts of the same large sheet of flagstone lying upon a mass of amphibolite and mica schist. Both *Holmsen,* who has recently been working south of Gjevilvatnet, and

Muret, who did reconnaissance studies throughout Trollheimen and areas to the south and west, confirm *Holtedahl's* preliminary interpretation of great recumbent nappes, similar to those of the Pennine Alps *(Strand* and *Holmsen,* 1960, p. 12; *Muret,* 1960, p. 31).

If the flagstone is actually repeated in the vicinity of Fosdals tjern as part of a great nappe, the "handfold" and the folds to the west must be synformal. One of the three cross sections in Fig. 26 has

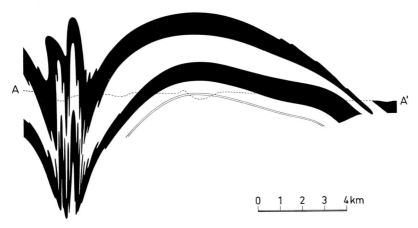

Fig. 48. Cross section reconstructed according to the synform hypothesis. Flagstone is shown in black. Location of the section is indicated on the geologic map (Pl. 1)

been reconstructed according to this hypothesis, and it is shown in Fig. 48. Though the existence of the folds may be due to the geometry of the nappe, they must owe their present configuration as synforms with a single *vertical* axis of symmetry for them all to a structural event that occured subsequent to the emplacement of the nappe and that so modified its structure as to obliterate it almost completely. The area of influence of the later event might be guessed to have extended as far as the northern slope of Mellomfjellet where the mineral lineation ceases to plunge toward Fosdals tjern and takes on the regional east-west trend, and as far in the other direction as the southern slopes of Gjevilvasskammene where the lineation also trends east-west; its influence east of Fosdals tjern is not readily determined by these means because both the regional trend and the imposed later trend of the lineation would roughly coincide in an east-west direction.

As in the salt-dome model, we are dealing with vertical tectonics for the formation of these structures, the stream lines being vertical at

Fosdals tjern, but the flow being directed relatively downward to produce synforms instead of upward to produce anticlines.

At this point, one might speculate upon the reason why this circular portion of the nappe should subside so much as to impose a new mineral lineation and other associated structures upon the rock. This part of the Caledonide chain is characterized not only by great recumbent folds, but also by large mantled domes of basement gneiss, which have apparently moved upward here and there and deformed the preexisting structures (*Strand* and *Holmsen*, 1960, pp. 12–13). One such dome, the Lønset anticline, lies about 20 kilometers south of Trollheimen (*Rosenqvist,* 1941). The Trollhetta granite gneiss as well as the quartzites and gneiss of the Gjevilvatn group have densities so low (ranging from 2.61 to 2.76) relative to those of the overlying Blåhø group (2.90 to 3.00) that they might possibly rise into domes under the influence of gravity during later stages of orogeny when they have gained enough mobility. Perhaps the homocline in the northern part of the area is the southern slope of such a dome, cored by Trollhetta granite gneiss, and perhaps a dome exists northwest of Storlifjellet. Even Svarthammer anticline, which bounds Riar basin on its southern and eastern sides, may owe its existence to such a process. The basin with its vertically attenuated synforms may have come into being when these domal anticlines began to rise and subtract material from beneath the area of Fosdals tjern to feed themselves. It may simply be a tight synform sucked down between larger mantled gneiss domes.

Choice of the Synform Hypothesis

For several reasons, the synform hypothesis appears the better of the two. Assigning the flagstone mass within the basin and on Storlifjellet to the same nappe to which the upper flagstone body on Hornet belongs simplifies the structure of the region. According to the anticline hypothesis, the configurations of the two masses are attributed to two different and unrelated processes — the formation of upright domes and the development of a recumbent nappe — the sphere of influence of one not being apparent in the sphere of the other. According to the synform hypothesis, the configurations of the two masses are both attributed to nappe formation, with the subsequent modification of one part of the nappe by basin development. Not only does this sequence of structural events seem simpler, but it has also been recognized in metamorphic cores of other mountain chains. Moreover, the synform hypothesis is a better fit to the local structure of Riar basin, which appears to be a simple basin imposed upon a conformable sequence (perhaps previously folded isoclinally and repeated by faults), without any indication that its central portion is antiformal or that a relatively open, rim syncline exists. Finally, the improbable shapes that the flagstone folds within the basin would have if they were indeed anticlines argue against the anticline hypothesis (Figs. 46, 47).

In conclusion, therefore, the flagstone mass that includes the "handfold" is considered here to represent part of a recumbent nappe, and the basin is considered to have developed by subsidence of its central portions relative to its periphery, preserving a portion of the nappe, now intensely refolded, at its center.

Flow Environment of the Sahlfold Facies

Sahlfolds, the mineral lineation that roughly parallels sahlfold axes, the elongate augen that parallel the mineral lineation and sahlfold axes, the schistosity that parallels sahlfold axial surfaces, and boudinage, the axes of which tend to be perpendicular to the mineral lineation and sahlfold axes, compose an assemblage of mesoscopic structures that have formed during the same period of time, as indicated by their relative spatial relationships and by their lack of mutual interference throughout Riar basin. This is the assemblage that is referred to collectively as the sahlfold facies. Enough of the local structural relationships have now been examined to permit an analysis of the type of flow that the sahlfold facies has recorded in the basin.

Essential Characteristics of the Flow

The slip by which slip folds develop can be described as inhomogeneous velocity gradient flow, in which the principal stream surfaces, df, parallel the axial surfaces of the folds (cf. Fig. 25). In accordance with the conclusion reached previously that sahlfolds were produced by slip, sahlfolds record velocity gradient flow across their axial surfaces, which parallel df. The stream lines, f, can be identified as the sahlfold slip lines, of which orientations were deduced in areas of tubular folds. The fact that the slip-line orientations within areas A, B, and C, as well as within much of the rest of the basin, converge downward both parallel with and across the sahlfold axial surfaces demonstrates either convergent, velocity gradient flow (c^2g) directed downward, or divergent, velocity gradient flow (d^2g) directed upward.

For the following reasons, structures of the sahlfold facies within the basin area are considered to have developed contemporaneously with the basin: (1) Although the form of the outer part of the basin is revealed by attitudes of the contacts between rock units, the form of its inner part is not obvious from the contacts because they reflect an earlier structural event as well as basining (Pl. 1; Fig. 26). Attitudes of the mineral lineation of the sahlfold facies, however, show the form of the whole basin (Fig. 27), as do the axes and slip-line orientations of sahlfolds. This spatial correlation of linear elements of the sahlfold facies with the basin's form implies a time correlation of the development of both the facies and the basin. (2) Basin formation was the last important structural event to occur in that part of Trollheimen. Structures of the sahlfold facies, though the norfold facies is superimposed upon them, are the last set of *intense* mesoscopic structures found in rocks within the basin. It seems reasonable to conclude that the last important structural event and the development of the last intense mesoscopic structures in the same area were contemporaneous. (3) Sahlfolds have been recognized in Trollheimen outside Riar basin (on western Høghø, for example) as have other folds of similar but slightly different aspects. During the field work, all such folds were classed as sahlfolds, though distinct differences were noted (cf. Chapter VIII), and in Plate 1, the folds are recorded by sahlfold symbols. Most of them do not correspond precisely to the sahlfolds described in Table 4, however, and until they are observed in enough detail for careful classification, they are considered to be fold types other than sahlfolds. In the western part of the area covered by Plate 1, structures strictly classed in the sahlfold facies are seen only in Riar basin, where the intensity of development of the facies increases to the basin center. This relationship further indicates contemporaneity of the basin and the facies.

Sahlfold slip lines within the basin plunge steeply inward toward the vicinity of the basin's center, as deduced from areas of tubular folds. The direction of mass transport by which the basin subsided relative to its surroundings must similarly be steep, presumably vertical at its center. From the contemporaneous development of the sahlfold facies and subsidence of the basin and from the similarities in orientation of sahlfold slip lines deduced in the basin and the direction of mass transport expected for the basin, it follows that the flow recorded by the sahlfold facies is at least part of the flow by which the basin subsided. Therefore, slip along the sahlfold slip lines was directed downward relative to the surrounding rock, and the flow was convergent, velocity gradient flow, c^2g, directed downward.

The preference of boudin axes for an orientation at high angles, commonly perpendicular, to the mineral lineation and sahlfold axes, and therefore at high angles to f, is compatible with convergent flow

of type c^2g, which requires shortening parallel to d and e and elongation parallel to f. On the other hand, this preferred orientation is
not compatible with the alternative of upward-directed divergent
flow, d^2g, which requires elongation parallel to d and e and shortening parallel to f (cf. Fig. 25). Elongation of augen parallel with the
mineral lineation and sahlfold axes, and therefore nearly parallel
with f, is also compatible with convergent, but not divergent, flow.

Relative Orientations of Fabric and Kinematic Axes

These relationships fix the orientations of the sahlfold fabric
axes, a, b, and c, relative to the kinematic or flow axes, d, e, and f
(Fig. 49). After *Sander* (1926, p. 328) and *Weiss* (1955, p. 229), the
fold axis is defined as fabric b, the axial surface as ab, and its pole
as c. Sahlfold axial surfaces, ab, parallel the principal stream surfaces, df; therefore, their poles, c and e, are parallel. In the three

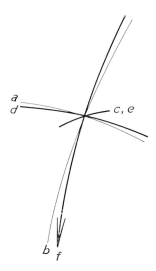

Fig. 49. Orientations of sahlfold fabric axes, a, b, and c relative to the principal
flow axes, d, e, and f, as determined within Riar basin

areas of tubular folds studied in detail, the average sahlfold axes, b,
are oriented 5° from parallelism with the stream lines, f. Thus, for
those folds, a is oriented 5° from parallelism with d. This general
relationship is maintained throughout the basin, though the angles
between a and d, b and f, may vary slightly.

It is noteworthy that the ac plane, which is a mirror plane in
cylindrical and cylindroidal folds like most sahlfolds, approximately

parallels the *de* plane, which is *not* a mirror plane in c²g flow (Table 1). This is merely an apparent discrepancy resulting from the different scales of observation of the fabric and the kinematics. Although sahlfolds in outcrop appear to be cylindrical and cylindroidal, the downward convergence of their axes throughout Riar basin indicates macroscopically that sahlfolds are conical. Similarly, although the slip lines as deduced are roughly parallel throughout any outcrop, thereby suggesting parallel flow (p²), the convergence of slip lines is recognizable macroscopically within groups of outcrops. Therefore, the *ac* and *de* planes are mesoscopically apparent, *but not real*, mirror planes.

Subordinate Characteristics of the Flow

During slip folding, a certain amount of differential slip must occur between particles within slip surfaces for the hinge lines of resulting slip folds to be of finite length (pp. 42—44); the same argument applies for the finite dimensions of their axial surfaces perpendicular to the slip lines. Thus, the fact that the axial surfaces of sahlfolds have finite dimensions indicates that a subordinate velocity gradient existed within *df*. In support of this conclusion one might argue that, in order for the basin to subside relative to its surroundings along stream surfaces that were not oriented tangent to the basin (cf. Figs. 35*a*, 37*a*, 38*a* with Pl. 1), some velocity gradient must have existed within the stream surfaces; flow in part of a surface near the center of the basin would presumably have been greater than flow in part of the same surface farther away from the center.

It is possible that the stream lines, rather than being perfectly straight, curved downward toward the vertical central axis of the basin (Fig. 50), and that the flow therefore was rotational. Although this in an appealing possibility, it would be necessary to deduce slip-line orientations at various depths in the basin to determine whether or not the stream lines were curved. Unfortunately, the land surface presents us with only a single level, so that the critical data are not available to us. If the curvature did exist, however, the flow would probably have been rotational about both *d* and *e* because, as was reasoned previously, the principal stream surfaces were not oriented tangent to the basin.

Any linear fabric element that is not oriented parallel with *f* would tend to be rotated relative to the flow axes by this type of flow. For example, a boudin line parallel with *d* would tend to be rotated out of parallelism with *d* toward *f* by rotation about *e*. Therefore, a systematic study of boudin lines relative to slip lines throughout the basin might reveal evidence of such rotation. Perhaps the boudins in which the modes of angles between boudin lines and the mineral lineation are significantly less than $90°$ can be explained in this way (Fig. 28b).

In summary, therefore, convergent velocity gradient flow of type c^2g is recorded by the sahlfold facies in Riar basin, though it might possibly be as complex as $c^2g^2r^2$.

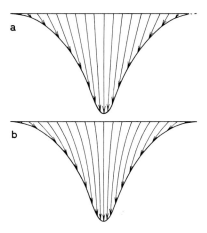

Fig. 50. Hypothetical, generalized cross sections of Riar basin during its development. a Subsidence along straight, converging stream lines. b Subsidence along curved, converging stream lines

Remarks on the Mechanism of Sahlfold Development

The model for sahlfold development is *constriction-slip folding*, which can be thought of as the nonaffine slip of plates that are simultaneously changing shape by constriction. This is c^2g flow in which g is nonlinear. It is assumed that the strain was mesoscopically triaxial, such that the plates lengthened parallel to f, shortened parallel to d, and shortened more parallel to e (i.e. $c_d < c_e$; analogous to paths of the deformation ellipsoid $1 < k < \infty$, *Flinn*, 1962; ellipsoid 4, *Ramsay*, 1967, pp. 162–165). This assumption is based both upon the greater probability that c_d did not equal c_e than that it did and upon the spread of linear fabric elements of sahlfolds toward the plane that represents the average attitude of sahlfold axial surfaces ($\sim df$), which suggests $c_d < c_e$.

It is interesting at this point to attempt a rough evaluation of the strain involved in, and the consequent rotation of sahlfold axes during, the flow. As estimated from the reconstructed cross section in

Fig. 48, the distance of flow of the rocks exposed in the flagstone "thumb" is 3 kilometers, assuming straight stream lines (Fig. 50*a*). An initial sphere of incompressible flagstone would be transformed by c^2 flow (taking $c_d = c_e$, for simplicity) of $1°$ across 40 meters for 3 kilometers into a prolate spheroid *(k = ∞, Flinn, 1962)*, of which the minor axis would be 4/9 the diameter of the original sphere and of which the ratio of major to minor axes would be about 12. A probable curvature of the stream lines as drawn in Fig. 50*b* requires the convergence during most of the flow to have been greater than the convergence now observed, and ignoring the consequent rotation, the resulting spheroid would be even more prolate than the one just described.

The mean attitudes of sahlfold axes at present are $5°$ from the mean stream-line orientations recorded within the flagstone "thumb." An estimate of the amount of rotation of such a fold axis can be obtained by assuming the folds to have formed at the initiation of flow and to have been carried 3 kilometers along straight stream lines converging at $1°$ across 40 meters (again taking $c_d = c_e$, for simplicity). The resulting value for the rotation is approximately $30°$. At the initiation of flow, therefore, the mean attitudes of sahlfold axes would have been approximately $25°$ from parallelism with the stream lines. Within the model of constriction-slip folding, however, individual folds would probably have begun developing at various times in the flow subsequent to flow initiation, and rotation of the fold axes would have been less than $30°$.

Certain aspects of sahlfolds, especially where they interfere with earlier folds, indicate that the layering may have had some strength during part of the development of the folds (p. 83). This possibility suggests an alternative model for sahlfold development, namely that sahlfolds formed by the flexural-slip mechanism and were subsequently altered to their present geometry by constriction during convergent flow. This model is not favored at present for two reasons: (1) It is not compatible with the profile geometry of sahlfolds, which are similar rather than flattened parallel (Table 4, line 1). (2) The estimated constriction associated with the convergent flow determined for the flagstone "thumb" is not extreme enough to alter the geometry of flexural-slip folds to approximate the geometry of slip folds. The paths that early lineations describe on the limbs and hinges of sahlfolds are nearly planar and are oriented at low angles (about $10°$) to the sahlfold axes (cf. Figs. 36*a*, 37*d*, 39*a*). In order to produce these relationships by the homogeneous constriction of flexural-slip folds with early lineations rotated into small circle paths, the fold axes must lie initially at high angles ($>60°$) to f and the angular rotation must be large ($>60°$). Neither the value of about $30°$ obtained in the previous paragraph for the rotation of the mean sahlfold axes during the flow nor the value of $35°$ obtained for the angle between the mean sahlfold axes and the stream lines at flow initiation is large enough.

Nevertheless, these values are based upon both the distance of flow and the convergence in stream lines, and it can be argued that the former was greater than the distance estimated or that the latter was greater than the convergence deduced. Neither of these possibilities seems likely, however, because the much greater distance of flow required by this model implies that Riar basin is extremely deep, much more so than the attitudes of rock units and structures there indicate at present, and because, as pointed out in Chapter IV (pp. 57—58), the methods of deducing stream-line orientations used here are based upon that component of slip contributed only by the velocity gradient and can therefore be used accurately to deduce a convergence of stream lines.

VII. The Norfold Facies

Structures of the norfold facies are superimposed upon those of the sahlfold facies in Trollheimen. They developed during a late and probably final stage in the formation of Riar basin and the surrounding Svarthammer anticline. They do not appear in the eastern homocline.

The only part of southeastern Trollheimen in which macroscopic folds are demonstrably norfolds is the northern portion of Mellomfjellet, where the "wrist" of the flagstone "handfold" joins the flagstone mass of Storlifjellet to the west (Pl. 1). There, the sharp turns of the flagstone unit and of the schist-amphibolite syncline east of it, from a northward trend to a westward trend, are wholly, or for the most part, macroscopic norfolds. No others are known.

Norfolds

A description of norfolds exposed in flagstone is given in Table 6, supplemented by Plates 12 through 16 and Fig. 51. In general these folds are uniformly open instead of isoclinal, and they are similar and harmonic in profile geometry, with fairly straight hinge lines and a moderate degree of cylindricity. Their limbs are planar, and thus the curvature of the folds is concentrated in their hinges, which are relatively sharp. Some of the names that are commonly applied to norfolds, especially to those in schistose rocks, are "kink fold," "crenulate fold," "chevron fold," "step fold," and "slip fold."

The appearance of norfolds is considerably more dependent upon the rock type in which the folds occur than is the appearance of either sahlfolds or discfolds. Three effects are noticeable:

1. Norfolds are larger in thick, massive layers than in thin, finely foliated layers. The difference in absolute linear dimensions is commonly greater than an order of magnitude and has been seen as great as two orders of magnitude. This effect may be the cause of multiple orders of norfolds in interlayered sequences. A thick, massive layer generally displays a single order of large norfolds, but the thin, nonmassive layers display two distinct orders — both the large ones seen in the massive layers, and presumably extended from those layers, and a set of smaller or higher-order norfolds peculiar to the nonmassive layers (Fig. 52).

Table 6. *Properties of norfolds in quartz schist (flagstone), Indre Kam formation*

Properties of an individual

1. Type of geometry	Similar (Pl. 14)
2. Nature of hinge and limbs	Small[a] to moderate radii of curvature, straight[a] or broadly curved limbs (Pl. 12). (Chevron-shaped[a] in highly schistose rocks; Pls. 13, 14. May resemble broad undulations in massive rocks; Pl. 15.)
3. Ratio of height to width	Range from 0.1 to 0.8 (cf. Fig. 4), based on $n = 27$. (Finely foliated amphibolite: 0.3 to 1.3 of $n = 50$.)
4. Ratio of depth to width	Large[b]
5. Length and character of hinge line	Short to long, and straight with some curvature (Pl. 12, bottom)
6. Cylindricity	Cylindroidal; some irregular (cf. Fig. 7)
7. Relation to cleavage	Foliation and sahlfold schistosity are folded (Pl. 16, bottom); poorly and locally developed (incipient) schistosity parallels axial surfaces. (Well developed slip cleavage parallels axial surfaces in highly schistose rocks[a]; Pl. 13; Pl. 14, bottom)
8. Relation to mineral lineation	Lineations within foliation and sahlfold schistosity are folded; poorly developed (incipient) lineation within foliation, parallel to fold axes, in some hinge areas

Properties of a group

9. Mean and standard deviation of height-width ratios	Mean of separate group = 0.4 of $n = 15$ (Fig. 51*a*); composite mean = 0.5 of $n = 27$ (Fig. 51*b*). Standard deviations = 0.21, 0.18[a] respectively. (Finely foliated amphibolite: means of separate groups = 0.5 of $n = 10$, 0.8 of $n = 10$, 0.4 of $n = 13$; composite mean = 0.6 of $n = 50$. Standard deviations = 0.14, 0.28, 0.19, respectively; composite standard deviation = 0.22[a])
10. Mean of depth-width ratios	Large[b]
11. Preferred orientation of fold axes	Linear distribution
12. Asymmetry	Normal (Pl. 14), with very large separation angle (150°+); some reverse on one limb and normal on opposite limb (Pl. 13, bottom; cf. Fig. 22*c*, *d*, *e*)

[a] Diagnostic property, useful in distinguishing between different types of folds in Trollheimen.

[b] Complete depths of norfolds have not been observed in uniform flagstone. However, incomplete depths have been seen to be a minimum of 10 times the widths of norfolds, and many are at least 20 times the widths. Thus, the depth-width ratios are concluded to be large. (In interlayered rocks such as alternating amphibolite and mica schist, the depths of norfolds are commonly terminated at sharp changes in rock type. This is apparently a function of the relative behavior of the different materials, rather than a property of the norfold disturbances, themselves. Therefore, only minimum depths of norfolds in uniform amphibolite and mica schist have been observed, and the resulting depth-width ratios are comparable to those in flagstone.)

Pl. 12. Norfolds in flagstone on northwestern Mellomfjellet. *Top.* Profile section showing chevron shape. *Bottom.* View of hinge lines

Pl. 13. Norfolds. *Top.* Poorly developed cleavage banding parallel to axial surfaces of norfolds in feldspathic mica schist, Kamtjerndalen. *Bottom.* Multiple orders of norfolds showing normal asymmetry relations on left limbs and reverse asymmetry relations on right limbs (cf. Fig. 22c, d, e) in gneiss of the Svarthammer formation, northeastern Blåhø.

Pl. 14. Norfolds in amphibolitic schists, exhibiting similar geometry, chevron shapes, and well developed cleavage parallel to axial surfaces. *Top.* Kamtjerndalen. *Bottom.* Northwestern Blåhø, within area of Fig. 53

Pl. 15. Norfolds superimposed upon sahlfolds in interlayered amphibolitic and feldspathic schist. *Top*. Northwestern Blåhø, within area of Fig. 53. *Bottom*. Western Blåhø, near Lille Kamtjern

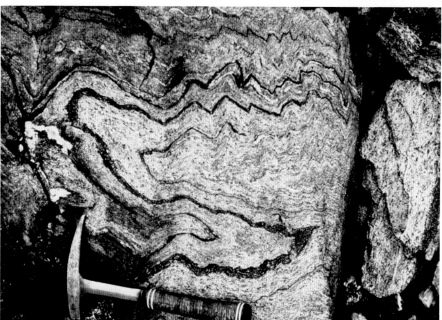

Pl. 16. Norfolds superimposed upon sahlfolds on northeastern Mellomfjellet. *Top.*
Norfold hinge lines wrapped around a sahlfold hinge in flagstone (cf. Fig. 21*b* and
discussion of the hinge-line node). *Bottom.* Feldspathic epidote amphibolite

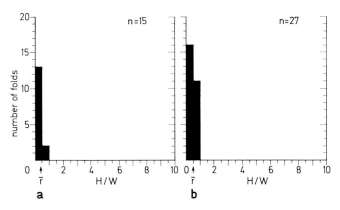

Fig. 51. Histograms showing height-width ratios of norfolds in flagstone. The mean values are indicated by arrows labelled $\bar{r}$ *a* Measured in a single outcrop on northeastern Blåhø. *b* Composite histogram of all the ratios measured in southeastern Trollheimen

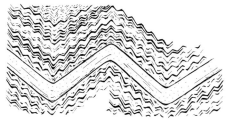

Fig. 52. Profile sketch of two orders of norfolds developed in layers of different composition

2. Within a sequence of interlayered rock types displaying nor-folds, certain thick layers of massive rock are not folded. Norfold axial surfaces are truncated at the contacts with such layers. The strain induced in these rocks by the stress that produced norfolds in the surrounding rocks was therefore either completely recoverable or nonrecoverable but homogeneous. This would be the extreme effect of rock type upon the appearance — or the very existence — of norfolds.

3. The hinges of norfolds are rounded in moderately massive rock types but sharp in highly schistose rocks. In addition, the weakly developed schistosity that parallels norfold axial surfaces in the more massive rocks is replaced by a slip cleavage in the schistose rocks. These differences in style combined with the difference in size may so change the appearance of norfolds from one rock type to another that the folds are sometimes mistaken for separate types. Neverthe-

less, the differences as described are strictly associated with changes in rock type and layer thickness; gradations in size and style occur as gradations in rock type and thickness occur. Furthermore, the remaining style criteria identify these folds as a single, if somewhat complex, fold type. Identities in orientations of norfold axial surfaces and fold axes in different rock types throughout the area support this conclusion.

Spatial Relations of Norfolds in Trollheimen

Axial surfaces of norfolds are commonly not parallel in any given outcrop. To determine how important the spread of attitudes might be, orientations of axial surfaces were measured throughout an area

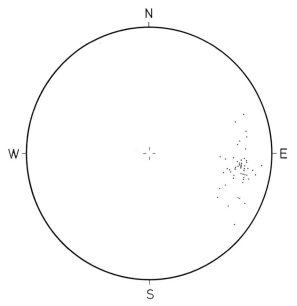

Fig. 53. Poles to 53 attitudes of norfold axial surfaces measured in an outcrop area on western Blåhø

where the spread was thought to be extreme. The area, 13,000 square meters in size, is located on western Blåhø and contains a variety of mica schists and amphibolites. Measurements of the axial surfaces (53 total) were taken with an effort to obtain the common orientations as well as the orientations that appear anomalous. Their poles, plotted in Fig. 53, define a point maximum that spreads equally in both directions toward a great-circle girdle. Seventy-five percent of the poles are oriented 15° or less from the center of gravity of the point maximum, and the spread of all the poles is 60°. Generally speaking, therefore, most of the axial surfaces in the area are parallel, and those that are not depart from the common, planar orientation in such a way as to maintain a common axis (π, the pole to the great-circle girdle of poles; *Turner* and *Weiss*, 1963, p. 83). Attitudes of norfold axial surfaces measured in other parts of Trollheimen show the same essential features.

Configuration of Axial Surfaces

A single attitude taken from the common attitudes of axial surfaces within an outcrop area is used here to represent that area. The size of the strong point maximum in Fig. 53 serves as a measure of the reliability of such an attitude within a similar distribution. Representative attitudes of norfold axial surfaces were obtained from 38 areas within Riar basin and Svarthammer anticline. They are shown on the map in Fig. 54, except for one that was measured on Storlifjellet, west of the map area (cf. Pl. 1).

The axial-surface attitudes recorded in Fig. 54 can be assigned to two major groups that effectively correspond to the major structural divisions of Svarthammer anticline and Riar basin. The dividing line (dashed) is near the contact between mica schist and amphibolite of the Kam Tjern formation and flagstone of the Indre Kam formation that crosses the prongs of Gjevilvasskammene, surrounds Blåhø to the south, east, and northeast, and trends northwestward on Mellomfjellet. Six attitudes of the axial surfaces are plotted on the map in Fig. 54 along this contact — one on Hemre Kam, two on Blåhø, and three on Mellomfjellet. The one on Hemre Kam (area 8) is anomalous in orientation and cannot be assigned to either group; the two on Blåhø and the southern two on Mellomfjellet are assigned to the anticline group; and the northernmost one on Mellomfjellet to the basin group. All attitudes east of this contact belong to the anticline group, which comprises 12 in all. All attitudes west of this contact belong to the basin group, which comprises 25 in all.

Axial surfaces in the anticline group describe an arch whose crest is located where the crest of Svarthammer anticline, as described by the foliation, is located, but whose axis has a consistently different plunge from that of the axis in the foliation. On Svarthammeren, where the fold axis in the foliation plunges about 15° to the southeast, the axis in the norfold axial surfaces plunges 40° or more in the same direction. In the central and southern portions of Falkfangerhø, where the axis in the foliation passes through the

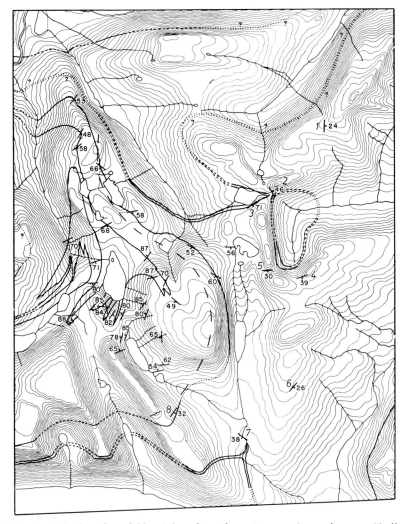

Fig. 54. Attitudes of norfold axial surfaces from 38 areas in southeastern Trollheimen (western two-thirds of Pl. 1)

horizontal and plunges about 10° to the north, the axis in the axial
surfaces plunges about 30° to the south. The remaining areas of norfold axial surfaces included in the anticline group, on eastern Mellomfjellet and Svarthammeren, show the breadth of the axial-surface
arch in its northern extension. Not enough observations were made on
foliation within this portion of the anticline to compare fold axes,
but the general broad foliation dome that appears to center near

Gjeithetta or somewhere to the north may also be coaxial with the axial-surface arch. Perhaps a late-stage rising of the Trollhetta granite to the north of this area has reversed the plunge of the northern extension of Svarthammer anticline, sharpened its hinges, and imposed on it the norfold axial-surface arch. This model has previously been suggested for "cleavage arches" associated with mantled gneiss domes elsewhere in the world (e. g. *White* and *Jahns,* 1950, pp. 214—219).

Axial surfaces of norfolds in the basin group are more irregular than in the anticline group. For example, norfolds are poorly developed between the "fingers" of the flagstone "handfold" on Riaren, and their axial surfaces vary greatly in attitude. Nevertheless, if such irregularities are ignored, a rough symmetry becomes apparent. In general, the axial surfaces strike northeastward throughout the basin, but their dip in the northern part is to the southeast and in the southern part to the northwest. If one were to traverse the basin from north to south, the axial surfaces could be seen to steepen gradually from northern Mellomfjellet, where they display moderate southeastward dips, to Fosdals tjern, where they are vertical, and then gradually to become shallow again in the valley of Kam tjern, where they display moderate northwestward dips. The locus of vertical axial surfaces — the vertical plane of symmetry — coincides with the northeastward trending fourth "finger" of the "handfold," the "palm" at Fosdals tjern, and the head of Slettå valley. It passes directly through the vertical center of the basin described by the foliation, the mineral lineation, and the axes of macroscopic and mesoscopic sahlfolds. Moreover, comparison of the map of norfold axial surfaces (Fig. 54) with the map on which the angle of plunge of the mineral lineation is contoured (Fig. 27) reveals the coincidence of the locus of vertical norfold axial surfaces and the axis of the "culmination" that localizes the identations in the lineation-plunge isogons.

It is concluded from the coincidence in location of the arch and fan of norfold axial surfaces with Svarthammer anticline and Riar basin, respectively, that norfolds formed in this region during the growth of the anticline and basin. Moreover, norfolds formed late in the development of these two structures, as shown by their superposition upon sahlfolds within the basin and on the basinward flank of the anticline (Pls. 15, 16).

Comparison of Norfold Orientations with Earlier Structures

In Riar basin, where the compositional layering is contorted by several generations of folding, the fabric of sahlfolds provides the most simple and regular set of reference axes for norfolds. The area of norfold development, however, which includes the outer portions of Svarthammer anticline, is considerably larger than the area of sahlfold development associated with Riar basin. Therefore, beyond the boundary of sahlfold development, elements of a presahlfold fabric are used for reference.

Unfortunately the outer boundary of sahlfold development — the line that separates rocks in which sahlfolds are rare from those in which sahlfolds are absent — is difficult to locate on Svarthammer anticline. Furthermore, the elements of fabric used here for reference, namely axial-surface schistosities and mineral lineations, are readily confused with corresponding structures of a different generation, especially where their orientations are nearly the same, as along the eastern border of Riar basin. For these reasons, the eastern boundary of sahlfold development was not recognized in the field, and the separation of sahlfolds from presahlfold structures was not accomplished during the collecting of the field data. Nevertheless, a satisfactory separation of these structures emerges from the analysis that follows. Though we start by looking at norfolds relative to two sets of prenorfold structures intermixed, we are able ultimately to draw conclusions about norfolds relative to each set individually.

Axial Surfaces; the Asymmetry of Norfolds. The upper, ruled histogram in Fig. 55 presents the dihedral angle between schistosity

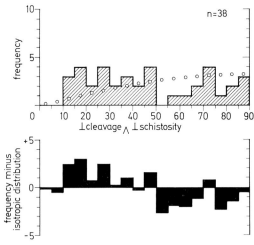

Fig. 55. Histogram showing the dihedral angle between pre-norfold schistosity and the norfold axial surface ("cleavage") in each of the 38 areas of Fig. 54. The dotted line represents the curve of the isotropic distribution of 38 linear elements on a sphere

and the norfold axial surface ("cleavage" in the figure) in each of
the 38 areas discussed in the preceding pages. Two striking features
of the histogram are the general absence of a mode and the lack
of angles between 0° and 10°.

Norfolds are difficult to see where their axial surfaces are inclined at 20° or less to schistosity, especially where the schistosity is
approximately parallel with the compositional layering. It seems
probable, therefore, that the lack of angles between 0° and 10° in the
histogram is the product of incomplete observation, rather than the
product of the mechanics of formation of norfolds themselves. Thus
this gap is ignored in considering the distribution in the histogram.

Because the poles to schistosity and to the norfold axial surfaces
are not restricted to a plane, but are free to occupy any and all
orientations in three dimensions, their correlation must be considered
by reference to the isotropic distribution of linear elements on a
sphere. This distribution is proportional to the sine of the angles
between the linear elements and a given reference axis (Bloss, 1957,
p. 218). The dotted line in the upper histogram of Fig. 55 represents
the isotropic distribution calculated for 5° intervals. The lower histogram shows the distribution in the upper histogram minus the isotropic
distribution. It is clear from the lower one that the distribution of
norfold axial surfaces relative to schistosity in the 38 areas is not
isotropic. With the exception of the interval from 0° to 10°, the histogram shows a direct correlation in the attitudes of norfold axial surfaces and schistosity in those areas.

Diagram a of Fig. 56 contains the contoured poles to schistosity
in the 38 areas, and diagram b contains the contoured poles to the
norfold axial surfaces. They are similar in their lack of strong point
maxima but dissimilar in their overall distributions. Diagram c combines a and b by presenting arrows that are drawn along great circles
in the short direction (<90°) from the poles to schistosity to the poles
to the norfold axial surfaces. (The lengths of these arrows represent
the dihedral angles plotted in Fig. 55.) Most of the arrows are long
and tend to parallel the periphery of the diagram, indicating that the
schistosity and the axial surfaces in most of the areas have different
strikes but nearly the same dip.

The direction in which an arrow such as these points relative to
the present vertical axis determines the asymmetry of an "average"
norfold exhibited by the schistosity within a given area (cf. Fig. 10).
The seven arrows clustered near the center of the diagram represent
the southernmost and easternmost areas on Svarthammer anticline
(areas 1, 2, 4—8, Fig. 54). They are both clockwise and counterclockwise, showing that the "average" norfolds in those areas display

both patterns of asymmetry. A comparison of Plate 1, Fig. 54, and the arrow diagram reveals that, in a sense, these norfolds bear a normal asymmetry relationship to the anticline, though their axial surfaces and fold axes are arched around it.

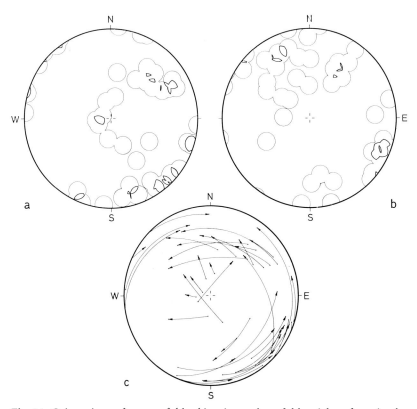

Fig. 56. Orientations of pre-norfold schistosity and norfold axial surfaces in the 38 areas of Fig. 54. *a* Poles to schistosity. Contours: 3, 8% per 1% area; maximum: 11%. *b* Poles to norfold axial surfaces. Contours: 3, 8, 13% per 1% area; maximum: 16%. *c* Great-circle arrows drawn in the short directions from schistosity poles to norfold axial-surface poles in corresponding areas

Of the remaining 31 arrows in the diagram, only 7 point in the clockwise direction. They represent areas within Riar basin, between the "fingers" of the flagstone "handfold" and to the west of the "handfold." The 24 counterclockwise arrows represent areas both within the basin and on the basinward flank of Svarthammer anticline, to the north, east, and south of the "handfold". The asymmetry of "average" norfolds in these 31 areas thus bears a systematic relationship to the center of the basin, as located by the "handfold". West of the "handfold" the asymmetry is counterclockwise; between the fingers the asymmetry is mixed; and north, east, and south of the "handfold" the asymmetry is clockwise. With the exception of the

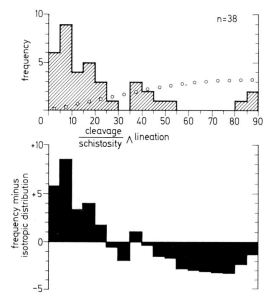

Fig. 57. Histogram showing the angle between the mineral lineation and the axis of an "average" norfold in each of the 38 areas of Fig. 54. The dotted line represents the curve of the isotropic distribution of 38 linear elements on a sphere

"thumb", where the asymmetry is clockwise on both limbs, norfolds exhibit a normal asymmetry relationship to the flagstone "handfold," though their axial surfaces do not parallel the axial surface(s) of the "handfold."

Fold Axes. The histogram in Fig. 57 presents the angle between the axis of an "average" norfold and the attitude of the pervasive mineral lineation in each of the 38 areas. The intersection of the norfold axial surface and the schistosity in a given area is taken as the "average" norfold axis ("cleavage/schistosity" in the figure). The curve of the isotropic distribution of linear elements on a sphere does not apply to the distribution in Fig. 57 because both the linear elements share a common plane, the schistosity. The histogram shows a strong direct correlation between the fold axes and the attitudes of the mineral lineation within schistosity.

9 Hansen, Strain Facies

The contoured diagrams in Fig. 58 contain the attitudes of the mineral lineation (*a*) and the "average" norfold axes (*b*) in the 38 areas. Diagram *a* shows a strong, vertical, point maximum and a weak, horizontal, east-west maximum. The 7 attitudes of the mineral lineation in the horizontal maximum represent areas 1 through 7

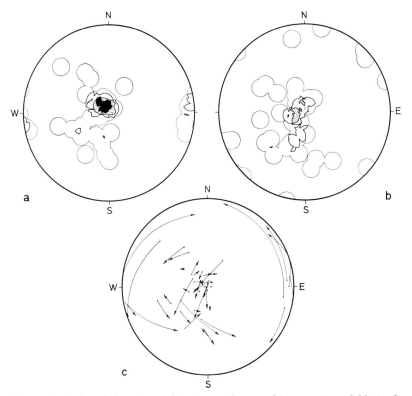

Fig. 58. Attitudes of the mineral lineation and axes of "average" norfolds in the 38 areas of Fig. 54. *a* Mineral lineation. Contours: 3, 8, 13, 18% per 1% area; maximum: 32%. *b* Norfold axes. Contours: 3, 8, 13% per 1% area; maximum: 18%. *c* Great-circle arrows drawn in the short directions from lineation attitudes to norfold axes in corresponding areas

on Svarthammer anticline (Fig. 54). Diagram *b* shows a single, nearly vertical, point maximum, weaker than the vertical maximum in *a*. The 7 fold axes within 20° of the periphery of the diagram represent areas 1 through 7.

Diagram *c* contains arrows (<90°) originating at the attitudes of the mineral lineation and pointing to the norfold axes for the 38 areas. A large number of the arrows are short, as expected from the distribution in Fig. 57. Arrows 15° and shorter are arbitrarily eliminated from consideration here because of their inherent high degree of uncertainty, which arises from the sampling of one attitude from a spread of attitudes of the mineral lineation, the schistosity, and the norfold axial surfaces in any single area (cf. Figs. 35a, c; 37a; 38a, c; 53); when this is done, 17 of the 38 arrow remain. Six of the 17 originate near the horizontal east-west position and represent the outer areas of Svarthammer anticline. They point both clockwise and counterclockwise relative to the present, vertical axis and appear to be nonsystematic. The remaining 11 arrows represent areas nearer the basin center, and they all point in a counterclockwise direction. Therefore, the axes of "average" norfolds to the north of the basin center plunge eastward of the mineral lineation, those to the east plunge southward of the mineral lineation, and those to the south plunge westward of the mineral lineation. Those to the west plunge southward of the mineral lineation, however, not northward, as might be expected from this pattern. The arrows from those areas pass nearly through the center of the diagram. It is expected, therefore, that additional data of this kind collected farther west of the basin center would yield clockwise arrows and that the norfold axes there would also plunge southward of the mineral lineation. It is interesting to note that, where the asymmetry of "average" norfolds is clockwise in the basin, the axes of the folds show a corresponding clockwise skewness from the attitude of the mineral lineation; where the asymmetry is counterclockwise, the axes show a counterclockwise skewness from the attitude of the mineral lineation. This correspondence is not governed by the geometry of the asymmetry of a fold in relation to the fold axis. It must be of regional significance.

Conclusions on Norfold Affinities. The isolation of attitudes of the mineral lineation in areas 1 through 7 from those in the other areas, and their sharing of the subhorizontal, east-west orientation (Fig. 58a), which is typical of the presahlfold mineral lineation throughout Trollheimen, lead to the conclusion that the lineation in areas 1 through 7 is presahlfold. The essential difference in character of the arrows representing these areas from those of the remaining areas in Fig. 58c is compatible with this conclusion.

The replacement or overprinting of one mineral lineation by another of a different orientation in metamorphic rocks is accomplished by recrystallization, at least for the most part, as is the overprinting of one schistosity by another. Since many of the same minerals contribute to those two structures, it is expected that the formation of a new mineral lineation in a rock proceeds concurrently with the development of a new schistosity if the controls are such that both structures will develop. It is further concluded, therefore, that the schistosity in areas 1 through 7 is also presahlfold. The differences in character of the arrows representing these areas from those of the other areas in Fig. 56c are compatible with this conclusion. Although the relationships in area 8 are anomalous, sahlfolds are well developed there; thus that area cannot be grouped with 1 through 7.

From the attitudes of the mineral lineation alone, the outer boundary of development of sahlfold-facies structures that accompanied basin formation can be located in the south somewhere on the prong of Midtre Kam (Pl. 1). On the basis of the arguments in the preceding paragraphs, the boundary extends eastward through Hemre Kam between areas 7 and 8, northeastward between Falkfangerhø and Blåhø, and northward just west of areas 5 and 3 (Fig. 54). Its location in the north in Slettå valley or on Gjeithetta is unknown, but it extends westward across the northern flank of Mellomfjellet at the valley of Folla. The zone of development of norfold-facies structures southeast and east of this sahlfold-facies boundary is of the order of 2 to 5 kilometers wide and includes areas 1 through 7 (Fig. 54).

The angles from this outer zone of norfolds that are included in the histograms of Figs. 55 and 57 are listed in Table 7. The angles in column ii encompass nearly the entire range of the distribution in Fig. 57 and show little indication of a mode. Thus the "average" norfold axes show no correlation with the presahlfold lineation in this zone. The angles in column i occur only in the lower half of the distribution in Fig. 55, thereby indicating a direct correlation between the attitudes of norfold axial surfaces and presahlfold schistosity in this outer zone.

Table 7. *Some angular relations between norfolds and earlier structural elements on outer Svarthammer anticline*

Area[a]	i $\left(\begin{array}{c}\perp \text{ norfold} \\ \text{axial surface}\end{array}\right) \wedge \text{schistosity}$	ii $\left(\begin{array}{c}\text{norfold} \\ \text{axial surface} \\ \text{schistosity}\end{array}\right) \wedge \left(\begin{array}{c}\text{mineral} \\ \text{lineation}\end{array}\right)$
1	11°	85°
2	37°	87°
3	48°	55°
4	23°	21°
5	12°	10°
6	19°	41°
7	50°	26°

[a] As numbered in Fig. 54.

After subtraction of the seven angles in column *i* of the table from the distribution in Fig. 55, the isotropic distribution curve is lowered by about 18 percent, and the attitudes of norfold axial surfaces show a weaker, but still obvious, direct correlation with the sahlfold schistosity in the remaining 31 areas. Subtraction of the seven angles in column *ii* from the distribution in Fig. 57 leaves a distribution that shows a strong, direct correlation between "average" norfold axes and attitudes of the sahlfold mineral lineation within sahlfold schistosity. These direct correlations among orientations of elements of norfold and sahlfold fabrics, both within the basin and on the basinward flank of the anticline, correspond on the scale of an outcrop to the correlation by location of the fan and arch of norfold axial surfaces with Riar basin and Svarthammer anticline, respectively, on the regional scale.

It was pointed out in the discussion of the arch and fan of norfold axial surfaces that the attitudes of those surfaces in 12 areas (the anticline group) appear to relate to the arch and that 25 of them (the basin group) appear to relate to the fan. (The attitude from area 8 is anomalous). Yet the area of development of sahlfold-facies structures, which relate only to the basin as shown in the preceding chapter, includes 31 of the areas, not 25. This difference in the affinities of the mesoscopic structures in 6 areas, from the time of development of the sahlfold facies to the time of development of the norfold facies, suggests a change in the regional importance of basin formation relative to anticline formation with time. Though merely an indication, it may mean that the formation of the basin involved a broader area in sahlfold time than in norfold time, and that the anticline grew at the areal expense of the basin during norfold time.

These relationships are compatible with the following model: Prior to sahlfold time, a somewhat flat, regional foliation with an east-west mineral lineation existed throughout most of Trollheimen. (For the present we can overlook the recumbent folds that also existed then.) Riar basin and Svarthammer anticline began to form more or less simultaneously upon this configuration. The process of basin formation was much more intense than the neighboring process of anticline formation, and the presahlfold fabric became overprinted by sahlfold-facies structures within the basin while it was merely up-warped passively on outer Svarthammer anticline. Deformation of the rocks involved in anticline formation increased as deformation of the rocks involved in basining decreased, and the less intense structures of the norfold facies developed in both regions. As the sequential product of the same, continuing, compound process that earlier produced sahlfolds, the norfold fabric was imprinted upon the sahlfold fabric with a direct correlation of elements. Outside the area of sahlfold development, however, and as the product of a process completely separate and different from the one that produced the presahlfold structures, the norfold fabric was imprinted upon the presahlfold fabric with a mixed correlation of elements. No correlation between norfold axes and the east-west mineral lineation resulted, as might be expected from their independent geneses, but a direct correlation of norfold axial surfaces and presahlfold schistosity did result, coincidentally, from the arching of the early, recumbent schistosity and the development of the arch of norfold axial surfaces, in the same rocks, both during anticline formation.

Having examined Trollheimen's norfolds in space and time, we now turn to the more specific problems of the mechanism of norfold development and norfold kinematics.

Rotation Paths of Early Lineations

The similar geometry displayed by norfolds in profile suggests slip folding as a tentative model of norfold development. The paths of early lineations rotated about norfolds can be used both to check the appropriateness of this model and, if appropriate, to determine the slip-line orientations that prevailed during norfold formation.

Rotated lineations were studied in 41 norfolds in 9 areas within Riar basin. The areas, D through L, vary in size up to 1000 square meters and occur around the flagstone "handfold" within the Kam Tjern formation. The rock types are amphibole mica schist and finely foliated amphibolite.

Two distinct mineral lineations were studied. One of them, displayed by large (up to 2 centimeters long), transversely fractured, amphibole needles, is presahlfold in age and was oriented initially at high angles to the imposed axes of norfolds. The other lineation, seen in small (about 5 millimeters long), unbroken, amphibole needles, as well as in most other minerals in the rocks, is the mineral lineation of the sahlfold facies and was oriented initially at relatively low angles to the norfold axes. (The histograms in Fig. 57 illustrate this relationship.) Both lineations are found in the same rock. For the present purpose, the presahlfold lineation is the more useful of the two because its rotation path is commonly longer and thus more diagnostic. Unfortunately, the presahlfold lineation is also more difficult to observe because it existed through the deformation and partial recrystallization that accompanied the development of the sahlfold facies in these rocks.

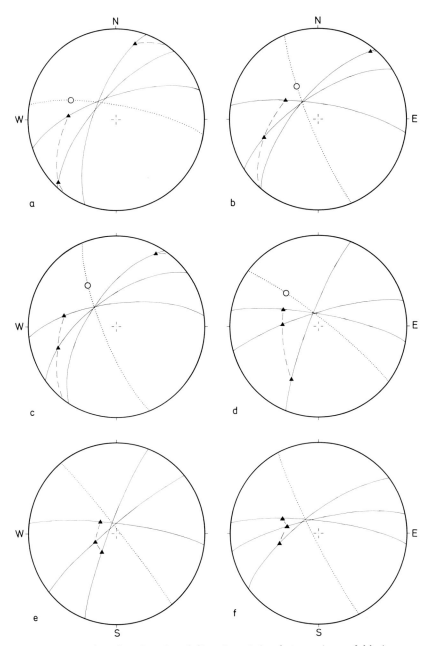

Fig. 59. Attitudes of early mineral lineations (triangles) on six norfolds in area F. Solid great circles: foliation. Dotted great circles: norfold axial surfaces. Dashed great-circle arcs: approximate continua of lineation attitudes. Open circles: slip-line solutions. *a, b,* and *c.* Folds 1, 2, and 3, respectively; pre-sahlfold lineation. *d, e,* and *f.* Folds 4, 5, and 6, respectively; sahlfold lineation

Six Folds in Area F

Area F is located on the northern slope of Hemre Kam, south of the "handfold." The norfolds studied there serve to illustrate the relationships found in norfolds in the nine areas.

Attitudes of early lineations measured on the limbs and hinges of six norfolds in this area are shown relative to the fold axes and axial-surface attitudes in Fig. 59. The lineation measured in folds 1, 2, and 3 is presahlfold (diagrams *a, b, c*), and the lineation in the remaining folds is sahlfold. Paths of the lineations in folds

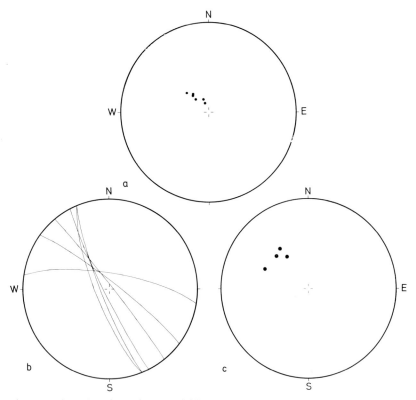

Fig. 60. Orientation data of six norfolds in area F. *a* Fold axes. *b* Axial surfaces. *c* Slip-line solutions

1 through 4 are roughly planar and can be approximated by great circles in the projections (Fig. 59*a*–*d*); this is compatible with the model of simple slip folding. The paths cannot be approximated by any of the possible small circles centered about the fold axes, as would be expected from a model of simple flexural-slip folding. The

lineation paths of folds 5 and 6 are not diagnostic (diagrams *e*, *f*); they are relatively short and could be fitted, though poorly, by either great circles or small circles.

The composite diagrams in Fig. 60 show (*a*) the six fold axes, (*b*) attitudes of the six axial surfaces, and (*c*) slip-line solutions by the Weiss method (pp. 27—28) for folds 1 through 4. Mutual intersections of the axial surfaces display a strong preferred orientation, which indicates the sharing of a common axis. Mean orientations of fold axes, intersections of axial surfaces, and slip-line solutions are not parallel, but their angular differences are small.

Summary of Relationships in Nine Areas

Certain critical angles of the norfolds examined in areas D through L are listed in Table 8. Paths of the early lineations rotated about 26 of the folds approximate planes like those in Fig. 59*a*—*d* and have been used to obtain solutions for slip-line orientations. Lineation paths in the remaining 15 folds are irregular like those in Fig. 59*e*, *f* and were not used for slip-line determinations; these folds are identified by the superscript e in the table. None of the lineation paths describes a small circle in spherical projection (i. e. has the same value in columns *iii*, *iv*, and *v* of Table 8a), but 6 of the 41 paths are moderately close (D3, D4, D5, G3, G4, H6). In summary, nearly $^2/_3$ of the folds examined display great-circle paths, $^1/_8$ display small-circle paths, and about $^1/_4$ display irregular paths.

Mechanisms of Norfold Development. It is clear from the foregoing tally that the rotation paths are not uniform enough to permit the conclusion that norfolds developed by slip folding. Therefore, in spite of the similar geometry of norfolds in profile, the question arises as to whether the folds might not be flexural-slip folds, of which some were subsequently flattened to produce the approximately planar lineation paths. This question can be answered in the negative for the following reasons: The average interlimb angle of the folds listed in Table 8 is 116° (column *i* of Table 8a). Thus, on the average, these are open folds that could not have been flattened enough to produce a great-circle rotation path from a small-circle path. Furthermore, the average interlimb angle of folds with great-circle paths is 119° ($n = 23$), that of the remaining folds is 110° ($n = 13$), and that of folds with small-circle paths is 103° ($n = 5$). Thus the folds that should have been flattened most, according to a flexural-slip flattening model, are more open on the average than those that should not have been flattened at all. This relationship is incompatible with flattening.

Table 8. *Angular relations of norfolds in nine areas of amphibolite and amphibole mica schist*

a) Within an individual

Area	Fold	i Inter-limb angle	ii $l'\wedge l'''^{a,b}$	iii f.a.$\wedge l'^a$	iv f.a.$\wedge l''^a$	v f.a.$\wedge l'''^{a,c}$	vi f.a.$\wedge$ sl.-line sol.
D	1	67°	139°d	46°	62°	132°	43°
	2	131°	113°d	52°	74°	161°	13°
	3e	86°	31°	18°	23°	22°	—
	4e	60°	47°	25°	31°	29°	—
	5e	94°	48°	35°	30°	38°	—
E	1	131°	44°d	29°	44°	59°	38°
	2	73°	40°	7°	13°	39°	21°
	3	137°	40°	33°	49°	62°	57°
F	1	133°	94°d	32°	87°	116°	26°
	2	124°	90°d	16°	47°	98°	16°
	3	134°	82°d	30°	49°	103°	20°
	4	103°	61°	28°	30°	64°	31°
	5e	105°	26°	12°	23°	26°	—
	6e	136°	21°	22°	18°	31°	—
G	1	133°	54°	37°	45°	73°	34°
	2	102°	54°	32°	40°	53°	42°
	3	140°	39°	69°	70°	72°	77°
	4	135°	44°	73°	72°	75°	86°
	5	129°	48°	61°	69°	77°	63°
	6e	76°	44°	17°	28°	37°	—
	7e	152°	13°	11°	8°	22°	—
H	1	110°	40°d	30°	26°	40°	43°
	2	118°	48°d	31°	45°	59°	54°
	3	—	56°d	30°	—	83°	8°
	4	—	29°d	27°	—	43°	40°
	5	—	15°d	46°	—	58°	30°
	6e	—	31°d	45°	47°	48°	—
	7e	127°	43°d	34°	68°	56°	—
I	1	126°	75°d	37°	57°	96°	37°
	2	114°	62°d	60°	69°	78°	72°
	3e	122°	57°	72°	76°	83°	—
J	1	91°	124°d	4°	85°	124°	21°
	2e	153°	15°	26°	23°	32°	—
	3e	116°	25°	23°	35°	25°	—
	4e	155°	20°	35°	39°	46°	—
K	1	144°	31°	23°	38°	45°	20°
	2	151°	31°	26°	37°	52°	27°
	3	88°	65°	31°	40°	57°	53°

Table 8. (*Continued*)

a) Within an individual

Area	Fold	i Inter-limb angle	ii $l' \wedge l'''^{a,b}$	iii f.a. $\wedge l'^a$	iv f.a. $\wedge l''^a$	v f.a. $\wedge l'''^{a,c}$	vi f.a. $\wedge$ sl.-line sol.
L	1	114°	27°	16°	19°	28°	33°
	2[e]	53°	60°	27°	30°	41°	—
	3[e]	—	25°	10°	33°	30°	—

b) Between mean orientations within an area[f]

Area	i f.a.'s $\wedge$ ax.-pl. X's	ii f.a.'s $\wedge$ sl.-line sols.	iii ax.-pl. $\wedge$ sl.-line X's sols.
D	3°	24°	27°
E	16°	38°	54°
F	9°	23°	14°
G	68°	58°	9°
H	7°	33°	34°
I	9°	54°	45°
J	15°	19°	10°
K	27°	31°	6°
L	—	30°	—

[a] l' = attitude of lineation on limb 1; l'' = attitude within hinge area; l''' = attitude on limb 2.

[b] This angle approximates the angle of rotation of the mineral lineation. It is measured along a *great circle* between corresponding ends of l' and l''', and includes l''.

[c] Measured between the same end of the fold axis and the corresponding end of l''' to those of l' and l'' that were used for columns *a-iii* and *a-iv*.

[d] Lineation formed by large, broken crystals of amphibole.

[e] Not used for determination of slip-line orientation.

[f] Means include axes and axial planes of all folds within an area, and not just of those folds used for slip-line determinations. All means are visual approximations.

It seems, therefore, that some combination of mechanisms might best explain the data at hand. A basis for such a combination can be found in the deformation history of the rocks. Prior to basin subsidence, deformation was apparently intense — probably the most intense in the history of the rocks — temperature and pressure were high, and recrystallization was total. During the early and main stage of basin development, when the sahlfold facies was imposed, temperature and pressure were apparently somewhat lower, recrystallization was important but not complete enough to obliterate the presahlfold schistosity and mineral lineation, and deformation was again intense. During a later stage of basin formation, when the norfold facies was imposed, temperature and pressure were lower still, recrystallization was unimportant except locally, and deformation was less intense. Therefore, two critical changes took place from sahlfold time through norfold time: (1) recrystallization declined from being a potential important means of effecting strain in the rocks to being inconsequential, and (2) owing to the lowering of temperature and pressure, rock strength increased and strength differences between individual rock types became more pronounced. Thus the environment of deformation gradually passed from one in which slip folding could take place in crystalline rocks toward one in which it could not and in which folding could be accomplished by flexural slip.

On the basis of this reasoning, it is concluded that norfolds developed initially by slip folding but subsequently and subordinately by some other mechanism — perhaps slip folding inhibited locally by "strong" layers or modified by external rotation in flexural slip. The geometry of individual norfolds should have recorded this compound origin, but this study was not detailed enough to reveal the expected evidence.

Mean Slip-Line Orientations. A mean orientation for the slip-line solutions in each area has been estimated visually for comparison with the regional structure of Riar basin. Before we proceed with the comparison, however, the spread among solutions within individual areas should be mentioned.

The largest angular difference between a solution and the mean in area F is 15°, and the average difference between all four solutions and the mean is 9° (Fig. 60c). The largest angular difference between a solution and a mean in any of the nine areas is 31°, and the largest average difference is 23°. Thus the angular differences among slip-line solutions for norfolds in these areas are large — much larger than those for sahlfolds in areas A, B, and C. Certainly part of this spread is lack of precision in measurement, but, in general, care was taken in the field to minimize this effect. Much of the spread is apparently due either to the primary differences in slip-line orientations during folding or to secondary rotations of some folds relative to others subsequent to their formation. The former seems as likely as the latter at present. Nevertheless, the mean slip-line orientations used in the following pages are based on few solutions with relatively large spreads, and whatever the cause of the spread, they may have large errors attached to them (like 30°), especially for the two areas in which only a single solution was obtained (J, L).

The only independent check on the accuracy of the mean slip-line orientations comes from area K. It uses the hinge-line node of a set of norfolds superimposed upon a sahlfold and is based upon the assumption that the norfolds developed by slip (cf. pp. 44—47, Fig. 21, and Pl. 16, *top*). Elements of the geometry are shown in Fig. 61a. Solid great circles represent attitudes of foliation about a sahlfold, triangles show the axes of norfolds within those planes, and the dashed great circle approximates the plane of norfold axes. The attitude of foliation that shows no disturbance by norfolding, and therefore contains the node in the norfold hinge lines, is shown

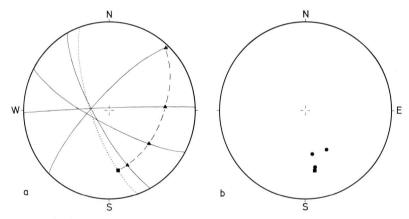

Fig. 61. Slip-line solutions for norfolds in area K. *a* Sahlfold with norfold axes superimposed. Solid great circles: foliation about the sahlfold. Triangles: norfold axes. Dashed great-circle arcs: approximate continuum of norfold axes. Dotted great circle: foliation containing norfold node. Square: slip-line solution. *b* Composite diagram comparing the solution from diagram *a* (square) with three solutions obtained by the Weiss method (circles)

as a dotted great circle. The intersection of the dashed and dotted great circles is a solution for the orientation of the norfold slip lines (cf. p. 46).

This solution is compared in Fig. 61b with three Weiss solutions obtained in the same area (Table 8). The angle between the single solution from the node and the mean of the other three solutions is 14°. Considering the spread already observed in slip-line solutions for norfolds, this value is acceptable; the solutions from the two different methods appear compatible.

Kinematic Significance of Fold Axes and Zone Axes of Axial Surfaces. The angle between the fold axis and the slip-line solution for each of the 26 norfolds displaying planar rotation paths is listed in column *vi* of Table 8a and shown in the histogram of Fig. 62. The angles spread through nearly the entire possible range, and their mode is

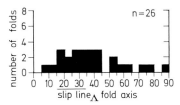

Fig. 62. Histogram of angles between fold axes and slip-line solutions of 26 norfolds

broad and occurs between 25° and 45°. Though the mode occurs in the low angles, it is considerably higher than that for sahlfolds (Fig. 43*d*). The angle between the mean of the fold axes and the mean slip-line orientation for each of the nine areas is listed in column *ii* of Table 8b. The angles spread through less than half of the entire possible range, and the mode is broad and occurs between 20° and 40°. Although an individual norfold axis has no apparent kinematic significance, the mean of norfold axes in an outcrop area occurs between 20° and 60° from the mean slip-line orientation for the area.

Mean orientations of the mutual intersections of axial surfaces in the nine areas show a slight direct correlation with the mean slip-line orientations, but they show a somewhat stronger direct correlation with the mean fold axes (columns *i, iii* of Table 8b). On the basis of these observations, the common axis of norfold axial surfaces within an outcrop area is of doubtful kinematic significance. Perhaps the orientations of these common axes are the products of a complex origin, such as primary development and selective, secondary rotation, which many of the other data on norfolds appear to reflect.

Regional Pattern of Slip-Line Orientations

Mean slip-line orientations for the nine areas are shown on the map in Fig. 63. The orientation for area L in the northeastern corner of the map is the only one obtained from an area on the eastern

limb of Svarthammer anticline, outside the basin. It plunges eastward, away from the crest of the anticline and away from the basin.

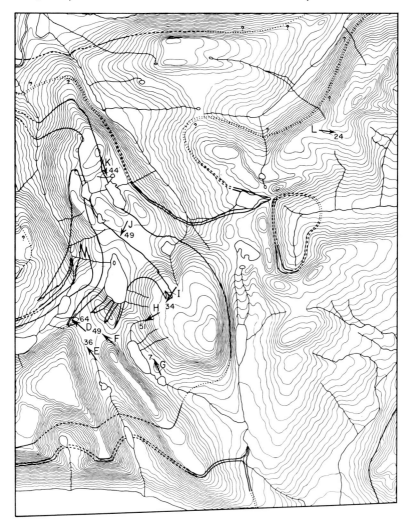

Fig. 63. Mean orientations of norfold slip lines in areas A through L in southeastern
Trollheimen (western two-thirds of Pl. 1)

Of the slip lines in the eight areas within the basin, those near
the center of the basin have steeper plunges than those more distant
from the center. Furthermore, two of them are radially directed relative to the basin center, and six are tangentially directed. Those

radially directed (G, J) plunge toward the center, similarly to the slip lines deduced from sahlfolds. Those tangentially directed (D, E, F, H, I, K) plunge in a clockwise direction relative to the center.

The common relationships among fabric elements of norfolds and sahlfolds within the basin to the north, east, and south of its center are summarized schematically in Fig. 64. Sahlfold axes (B′) and slip lines (sl′) within the sahlfold schistosity (S′) plunge toward the basin

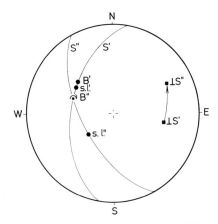

Fig. 64. Common relationships among norfold and sahlfold fabric elements in Riar basin. Single prime denotes sahlfold elements; double prime denotes norfold elements

center. Average norfold axes (B″) are slightly skewed from sahlfold axes in a counterclockwise direction; thus they plunge in a clockwise direction relative to the basin center. Norfold axial surfaces (S″) are displaced in a counterclockwise direction from sahlfold schistosity, and the norfold asymmetry is clockwise. Norfold slip lines (sl″) are skewed from norfold axes in a counterclockwise direction and plunge in a clockwise direction relative to the basin center. Though no slip-line solutions were obtained from norfolds to the west of the basin center, it is likely from the reversal of the other relationships described that norfold slip lines there plunge in a counterclockwise direction relative to the center.

It appears, therefore, that the prevalent slip-line orientations in the basin changed from radial in sahlfold time to tangential in nor-fold time. The radial solutions from norfolds in two of the outcrop areas suggest that the change either occurred gradually or did not occur everywhere.

Flow Environment of the Norfold Facies

The norfold facies is an assemblage of mesoscopic structures that comprises norfolds as well as the locally developed, incipient mineral lineation that parallels norfold axes and the local, incipient schistosity (or in highly schistose rocks, the well developed slip cleavage) that parallels norfold axial surfaces. Less is known about the norfold facies than the sahlfold facies at present because the norfold facies comprises fewer structures and therefore offers fewer relationships to analyze, and because the structures it does comprise are considerably less regular. Nevertheless, for the present purpose it is useful to draw conclusions, if only tentative, on the flow recorded by the norfold facies in Trollheimen.

The slip by which slip folds develop is described in previous chapters as inhomogeneous velocity gradient flow, in which the principal stream surfaces, df, parallel the axial surfaces of the folds. In accordance with the conclusion that the majority of norfolds in Trollheimen were produced by slip, those norfolds can be said to record velocity gradient flow across their axial surfaces, which parallel df. The norfold slip lines can be identified as stream lines, f. We do not know enough about the remaining (late?) norfolds to comment upon their relationships to df and f.

Restrictions from the Configuration of Riar Basin

It is possible that the rocks containing the norfolds from which the slip-line solutions were obtained were subsequently rotated relative to the basin center either during late norfolding or during some separate event. Nevertheless, no evidence for a penetrative structural event later than norfold development has been recognized in Trollheimen, and the rotation by later norfolding was probably local and minor, merely causing scatter among solutions within individual outcrop areas. Therefore, it is assumed that the stream-line (slip-line) orientations now recorded by norfolds are the same relative to the basin center as those that obtained during norfold development.

It is also assumed that flow along the norfold stream lines was downward in the basin relative to the surrounding rocks. That is to say that the norfold facies resulted from the continuing process of basin subsidence that earlier resulted in the sahlfold facies in the same rocks. To assume the reverse — that flow was directed upward — would be to assume that the basin ceased to subside relative to the anticline after sahlfold time and began to rise, or that the anticline began to subside. This possibility seems remote.

10 Hansen, Strain Facies

The sketches in Fig. 65 show the generalized attitudes of stream lines in Riar basin during sahlfold time (*a*) and norfold time (*b*). The downward converging, conical form of the sahlfold stream lines is also seen in the norfold stream lines, as evidenced by their not being perfectly tangential to the basin center but plunging slightly inward and by their having steeper plunges near the basin center (Fig. 63).

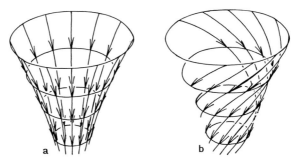

Fig. 65. Generalized attitudes of stream lines in Riar basin during sahlfold time *(a)* and norfold time *(b)*

The principal difference between the two forms is that the sahlfold stream lines plunge more or less straight down the cone but the norfold stream lines twist down the cone. The flow in both cases is convergent and rotational.

The quantitative convergence of norfold stream lines in the basin is not given here because its calculation requires more information about the form of the cone and the orientations of the stream lines and stream surfaces than is known at present.

Other Considerations

Boudinage or augen elongation that relates to norfolds has not been recognized in Trollheimen. This negative evidence suggests that the constrictional strain involved in norfold development was much less than that involved in sahlfold development. Either the convergence of norfold stream lines was less or the distance of flow along the stream lines was shorter. The style of norfolds is compatible with this conclusion; the short, curved hinge lines and the low degree of cylindricity indicate little or no stretching of the folds along their axes, which, on the average, are oriented about $30°$ from *f*, the direction of maximum elongation in convergent flow. It is apparent from these considerations that constrictional strain may not be important to the development of the norfold facies.

The discontinuous development of norfolds within the basin also permits this possibility. Norfolds commonly are well developed in one rock type but absent in an adjacent one, and they are absent in outcrops of some rock types in which they are well developed elsewhere. Since norfolds appear to occur in isolated units of rock, it follows that the flow in which the folds resulted may also have occurred in isolated units of rock. It is possible, therefore, for the flow within volumes of rock represented by outcrops or groups of outcrops to have been somewhat different from the overall flow in the basin; the mesoscopic flow may have been controlled by, but not identical to, the macroscopic flow. For example, one can imagine discontinuous subvolumes of rock undergoing parallel flow but arranged in such a way as to accomplish convergent flow on the scale of the basin. (The radially and tangentially oriented slip lines in the basin may have resulted in this manner contemporaneously, rather than at different times in a changing flow regime, as suggested earlier.) These speculations can be concluded with the hypothesis, based upon the pattern of norfold stream lines in the basin (Figs. 64, 65b) and upon the considerations in the foregoing paragraph, that mesoscopically the stream lines converged with respect to the e axis of flow but were approximately parallel within df; the flow would be pc instead of c^2. A subordinate velocity gradient within df during flow could account for the relatively short hinge lines of the norfolds that developed by slip (cf. pp. 42--44).

In summary, the flow recorded by the norfold facies in Trollheimen is poorly known at present, but the flow in which most of the norfolds in Riar basin developed is concluded provisionally to be c^2gr^2 or $pcgr^2$, though it may be as complex as $c^2g^2r^2$ or pcg^2r^2. The flow in the basin may have been spatially discontinuous, the stiffening rocks having moved along short, twisting paths to accomplish the final tightening of the basin.

10*

VIII. The Discfold Facies

The structure of the eastern portion of the area covered by Plate 1 can be described broadly as a homocline. Its dip is rather uniform to the east (Fig. 66), but its complexity varies drastically and discontinuously across strike. To the west it is relatively simple. As one walks from the crest of Svarthammer anticline on Falkfangerhø eastward to Tyrikvamfjellet and Høghø, one passes steadily upward in the rock column from flagstone of the Indre Kam formation into schist and amphibolite of the Kam Tjern formation. Throughout this western zone of the eastern homocline, the mineral lineation plunges

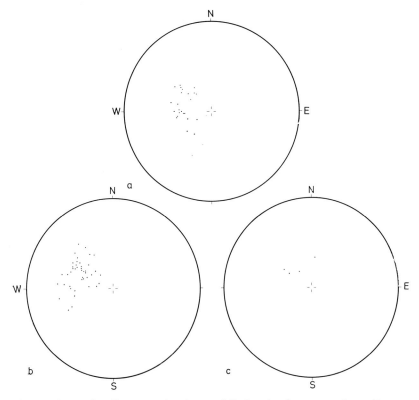

Fig. 66. Composite diagrams of poles to foliation in the eastern homocline. *a* Western zone, 24 poles. *b* Svahø border zone, 43 poles. *c* Trondheim basin, 4 poles

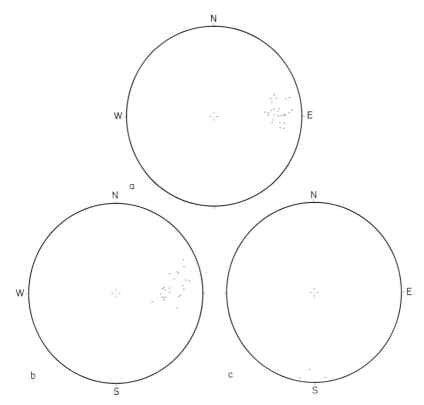

Fig. 67. Composite diagrams of attitudes of the mineral lineation in the eastern homocline. *a* Western zone, 28 attitudes. *b* Svahø border zone, 24 attitudes. *c* Trondheim basin, 3 attitudes.

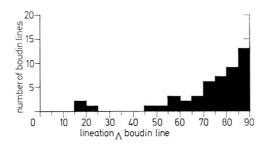

Fig. 68. Histogram showing the angular orientation of 50 boudin lines with respect to the local mineral lineation within mica schist and amphibolite on Tyrikvamfjellet

uniformly to the east (Fig. 67a). Mesoscopic folds are sparse, and with the exception of a macroscopic fold on Høghø (Pl. 1), the foliation is planar and relatively undisturbed. Boudinage in the schist-amphibolite sequence is abundant, and boudin lines tend to lie at right angles to the mineral lineation (Fig. 68).

Continuing eastward to Hammerhø and Grønhø, one passes back into rocks of the Gjevilvatn group, including thick sequences of flag-stone and augen gneiss with interfoliated micaceous marble, amphib-olite, and mica schist. These rocks have not been separated into the formations recognized farther west because the sequence is not quite the same and because the structure is not adequately understood. This relatively complex eastern part of the homocline contains the easternmost portions of the high-grade rocks of Trollheimen and is called the *Svahø border zone*. In the mapped area, the Svahø border zone comprises Svahø, Hammerhø, Grønhø, Skrikhø, and part of Nonshø (Pl. 1).

Structures of the discfold facies characterize the Svahø border zone. Discfolds are superimposed upon several earlier sets of folds and rotate the regional foliation and pervasive mineral lineation. They are the last folds to have been imprinted upon the rocks, and they appear to have developed during the emplacement of a large, recumbent nappe complex, comprising all the Svahø zone. Although postcrystallization structures for the most part, they are thought to have developed earlier than the sahlfolds and norfolds in the west because the discfold event of nappe emplacement occured prior to the sahlfold and norfold event of basin and anticline formation.

Macroscopic Structures of the Svahø Border Zone
Faults and the Zone Boundaries

The discontinuous nature of the eastern homocline, and of the Svahø zone in particular, is due to the presence of faults. In this respect, the eastern homocline stands in sharp contrast to Riar basin, which can be described completely by the geometry of folding. The relative importance of faults and folds in controlling the structure of the homocline is a matter of interpretation, however, because the fault surfaces tend to parallel the foliation, and their existence is difficult to prove.

The most profound of the faults is the *Nonshø border fault*, which parallels the eastern edge of the map (Pl. 1) for its whole north-south extent. The trace of the Nonshø fault on the land surface lies east of the mapped area, except in the southeastern corner of the map, where the fault trends slightly southwestward and intersects

Nonshø. The fault surface is not exposed, but rocks on the two sides of it are exposed less than 5 meters apart. Though the fault parallels regional foliation, mesoscopic structures are disharmonic to its surface and change their attitudes by nearly 90° from one side to the other (Figs. 67b, c; 72b, c). The fault separates greenschist with biotite and albite from amphibolite, marble, and mica schist with almandine and oligoclase, indicating that the highest greenschist subfacies has been omitted. (As bulk compositions suitable to the formation of kyanite or sillimanite are absent, it is possible that the lowest almandine-amphibolite subfacies has also been omitted.) Therefore, the Nonshø fault separates the low-grade rocks of the Trondheim basin on the east from the high-grade rocks of Trollheimen on the west; it is the eastern tectonic boundary of both Trollheimen and the Svahø border zone.

Evidence for a second fault west of the Nonshø fault is more equivocal. A fault surface is exposed on western Grønhø in an isolated outcrop where calcsilicate layers rest on calcareous schist with an angular discordance greater than 45°. Though the surface itself was not seen to the north or south along its strike, a major structural discordance occurs along its southward projection between Tyrikvamfjellet on the west and Hammerhø and Svahø on the east. Nevertheless, there are places where one can walk on continuous exposure from Høghø to Grønhø and from Tyrikvamfjellet to Hammerhø across the probable southward extension of the *Grønhø fault* without seeing any indication of its presence. If it exists at all in these areas, therefore, it must be either a foliation fault or a wide zone of foliation-shear surfaces (slide). Whatever its nature actually may be, or wherever it may be located (it is not drawn on Pl. 1), the Grønhø fault separates the fairly regular western zone of the homocline, composed of medium-grained, totally recrystallized rock, devoid of structures of the discfold facies, from the Svahø border zone, composed of fine-grained, dominantly cataclastic rock with abundant structures of the discfold facies. No discordance of mineral facies was recognized.

Within the Svahø zone, between the Grønhø and Nonshø faults, minor structural anomalies and the appearance and disappearance of rock units suggest additional fault control. Nevertheless, the zone has not been mapped in the detail required to delineate such structures there, if they do exist.

The Hammerhø Synform

The western face of Hammerhø near its junction with Svahø exposes the hinge area of a large recumbent synform. The radius of curvature in the hinge area is moderate, and the limbs are broadly curved. The fold is similar in profile geometry; its cylindricity is high. Compositional layering displays the fold, and a pervasive schistosity parallels its axial surface, which dips about 30° to the east. Its fold axis plunges 24° slightly north of east (Pl. 1). To the northeast on Hammerhø, structurally above the hinge area of the synform, the intersection of compositional layering and schistosity, which parallels the axis of the synform, changes gradually from the nearly east-west trend with shallow plunge to a northeast trend, horizontal (Pl. 1). The attitude of schistosity, which can be taken for the orientation of the axial surface there, dips to the east at about 30°. Therefore, most of Hammerhø is underlain by this macroscopic synform, of which the hinge line is at least 2 kilometers long, the axial orientations change through 30° or 40°, and the axial surface is shallow and somewhat planar.

This fold can be either a recumbent syncline or the inverted nose of a recumbent anticline or nappe. We do not know the sequence of rock units there well enough to decide between the two, but structural evidence discussed subsequently in this chapter suggests that the fold is actually an inverted anticline and that the Grønhø fault or slide may be its sheared lower limb. (The eastern end of the cross section in Fig. 48 is reconstructed according to this model.) It is further possible that nappe structure dominates the whole Svahø zone *(Holtedahl,* 1950, p. 56, Fig. 18), though the synform on Hammerhø is the only macroscopic, recumbent fold observed in the zone.

A System of Flexures

Foliation throughout the Svahø border zone is distorted locally into large, gentle flexures that lack a schistosity parallel to their axial surfaces. The relationship of one flexure to another is unlike the macroscopic folds in Riar basin, where fold axes vary systematically in direction and value of plunge from one locality to the next. Here, instead, the axis of one flexure may bear any relationship to the axis of another beside it, within the local attitude of the homocline. Their apparent lack of a significant effect upon the configuration of the rock units discouraged close mapping of these folds, but wherever encountered, their fold axes were determined as β's *(Turner* and *Weiss,* 1963, p. 154), and their asymmetry patterns were recorded. The equal-area projection in Fig. 69 shows eleven such axes from widely separated areas on Svahø, Hammerhø, Nonshø, and Grønhø. The fold axes display a planar preferred orientation that parallels the general attitude of foliation (Fig. 66b), and the asymmetry patterns define a separation angle of 18°.

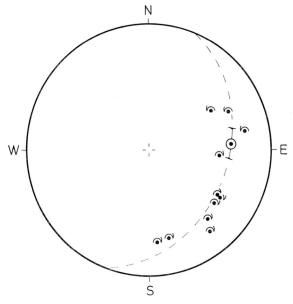

Fig. 69. Fold axes and asymmetry of 11 large flexures in foliation within the Svahø border zone. Separation angle, 18°

Mesoscopic Structures in the Svahø Border Zone

Rocks in the Svahø zone display several varieties and generations of medium-scale structures. Although most of them do not belong to the discfold facies, they are described briefly at this point because they offer insight into the general strain environment prior to disc-fold time and because, as passive markers, they contribute to the record of the discfold deformation.

The Regional Mineral Lineation

Everywhere west of the Svahø zone, with the exception of Riar basin, the pervasive mineral lineation has an east-west trend (Pl. 1). The same regional trend is exhibited by the lineation within the Svahø zone (Fig. 67b). The lineation-plunge isogonic map (Fig. 27) shows, however, that the plunge of the lineation is not so systematic or uniform within the Svahø zone as it is to the west. Attention has been called to the regularity with which the lineation in Riar basin steepens to a vertical central axis and to the broad, systematic sweep of the lineation on Svarthammer anticline. In comparison, the eastern homocline and especially the Svahø zone are seen in the plunge of the mineral lineation to be irregularly warped and buckled (Fig. 27).

Elongate Augen

Large masses as well as thin sheets of augen gneiss are found in the zone (Pl. 1). Ratios of augen dimensions commonly range up to 1:4:30. Long axes of the augen parallel the regional trend of the mineral lineation, and intermediate axes parallel the foliation.

Boudinage

Though boudins were not systematically observed in the Svahø zone, the orientations of boudin lines were noted wherever encountered. They exhibit nearly north-south strikes and shallow plunges, at high angles to the regional trend of the mineral lineation.

Hammerhø Synform Folds

A set of higher order folds are found on the limbs of the Hammerhø synform. They are coaxial with the synform and share the style characteristics and orientations described for the synform itself. They have been recognized only on Hammerhø.

Lang Tjern Folds

The folds at Lang tjern between Hammerhø and Nonshø are very harmonic, similar folds, reclined and isoclinal (Pl. 17). They are seen only in the compositional layering; the prevailing schistosity

Pl. 17. Lang tjern folds in flagstone at Lang tjern

Pl. 17

parallels their axial surfaces. Their hinge lines are long and straight, they display a high degree of cylindricity, and their fold axes display a linear preferred orientation that trends east-west and plunges shallowly to the east.

Pl. 18. Høghø folds in feldspathic quartzite on eastern Høghø, at junction with Grønhø

Høghø Folds

The folds on eastern Høghø at its junction with Grønhø are upright, similar folds, neither open nor isoclinal, with long, straight hinge lines and a high degree of cylindricity (Pl. 18). They rotate the foliation and display a schistosity parallel to their axial surfaces. Their fold axes show a strong, linear preferred orientation that trends east-west and plunges shallowly to the east, like that of the Lang tjern folds.

Interference Structures

Interference structures of Høghø folds on Lang tjern or Hammerhø synform folds were seen in flagstone in a small area (about 100 meters square) between Grønhø and Høghø. No other interference structures were found east of Riar basin.

The mesoscopic strain features described to this point are all considered prediscfold.

Discfolds

Discfolds exposed in flagstone within the Svahø zone are described in Table 9 and illustrated in Plates 19 and 20 and Figs. 70 to 72. Generally speaking, they are disharmonic, parallel folds, rather open on the average, with short, curved hinge lines and a low degree of cylindricity. Characteristically, there is neither a cleavage of any sort parallel to their axial surfaces nor a mineral lineation parallel to their fold axes. Their fold axes show planar preferred orientations, and the distributions of their asymmetry patterns define small separation arcs in spherical projection. Perhaps the only commonly used names for discfolds are "concentric folds" and "parallel folds", which are also applied to folds that are not discfolds.

In addition to their occurrence in the Svahø zone, discfolds have been found in flagstone and amphibolite in a few isolated localities near the southern edge of Riar basin (Pl. 1).

Mode of Discfold Development

Thickness of compositional layers within profile sections of discfolds in flagstone indicate that the folds are parallel but not ideally so. Compositional layers measured parallel with the axial surfaces are markedly thinner in the hinge areas than on the limbs; this is characteristic of parallel geometry. But the same layers measured perpendicular to the layer interfaces are thicker in the hinge areas than on the limbs; this is characteristic of similar geometry or of flattening (*Ramsay*, 1962a).

Table 9. *Properties of discfolds in quartz schist (flagstone), Indre Kam formation*

Properties of an individual

1. Type of geometry — Parallel[a] (Pl. 19)

2. Nature of hinge and limbs — Small to large[a] radii of curvature (Pl. 19); the two extremes are commonly displayed by a single layer, the radius within the concave surface of a folded layer being smaller than the radius within the convex surface (i. e. parallel geometry). Broadly curved limbs.

3. Ratio of height to width — Range from 0.1 to 1.0 (cf. Fig. 4), based on $n = 36$

4. Ratio of depth to width — Range from 0.6 to 5.3 (cf. Fig. 6), based on $n = 24$

5. Length and character of hinge line — Short and curved[a]

6. Cylindricity — Cylindroidal and irregular[a] (cf. Fig. 7)

7. Relation to cleavage — Foliation is folded (Pls. 19, 20); commonly no cleavage parallels axial surfaces[a], but a poorly and locally developed (incipient) schistosity parallels some axial surfaces

8. Relation to mineral lineation — Lineation within foliation is folded; commonly no lineation parallels fold axes, but a poorly developed (incipient) lineation within foliation, parallel with fold axes, is present in some hinge areas

Properties of a group

9. Mean and standard deviation of height-width ratios — Means of separate groups $= 0.3$ of $n = 12$, 0.5 of $n = 8$ (Fig. 70a, b); composite mean $= 0.5$ of $n = 36$ (Fig. 70c). Standard deviations $= 0.12$, 0.29 respectively; composite standard deviation $= 0.25$. (Finely foliated amphibolite: mean $= 0.8$ of $n = 13$; standard deviation $= 0.41$, Fig. 70d)

10. Mean of depth-width ratios — Composite mean $= 2.6$[a] of $n = 24$ (Fig. 71)

11. Preferred orientation of fold axes — Planar distribution[a] (Fig. 72a, b, d); some groups display linear distributions that spread toward planes (Fig. 72c)

12. Asymmetry — Normal; separation angles range from 0° to 150°, though small angles common (Fig. 72)

[a] Diagnostic property, useful in distinguishing between different types of folds in Trollheimen.

Pl. 19. Profile sections of discfolds in flagstone on Svahø

Pl. 20. Discfolds in flagstone on Svahø. *Top*. View showing barrel-shaped hinge. *Bottom*. Discfold with sharp hinge (near hammer) and irregular limb

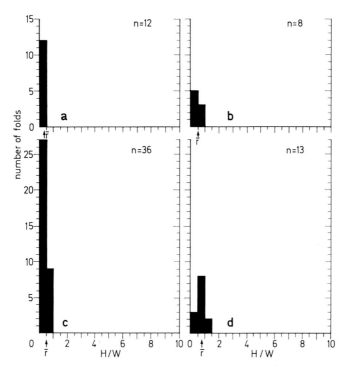

Fig. 70. Histograms showing height-width ratios of discfolds. Mean values are indicated by arrows labelled $\bar{r}$. *a* Measured in a single outcrop of flagstone on western Nonshø. *b* Measured in a single outcrop of flagstone on western Nonshø. *c* Composite histogram of all ratios measured in flagstone in the Svahø border zone. *d* Measured in a single outcrop of amphibolite on Hemre Kam

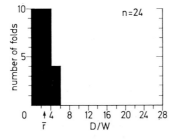

Fig. 71. Composite histogram of all depth-width ratios measured in flagstone in the Svahø border zone. The mean value is indicated by the arrow labelled $\bar{r}$

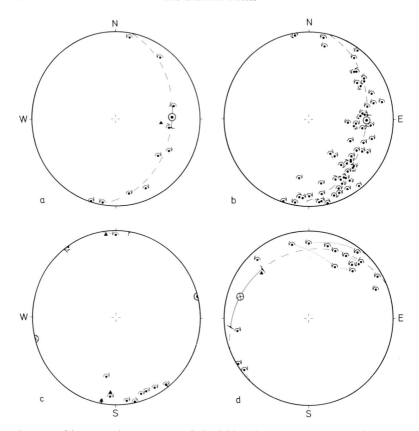

Fig. 72. Fold axes and asymmetry of discfolds. Planes approximating the spread of axes are drawn as great circles. Attitudes of mineral lineations shown by triangles. *a* 10 axes measured in flagstone on western Nonshø; 18° separation angle. *b* Composite diagram of 61 axes measured in various rocks of the Svahø border zone; 9° separation angle. *c* 10 axes measured in quartzite on eastern Nonshø; 133° separation angle. *d* 17 axes measured in amphibolite on Hemre Kam; 53° separation angle

Rotation paths of the regional, east-west lineation were observed in nine discfolds in flagstone on Svahø. Although one of them is a nearly perfect small-circle path centered about β (Fig. 73*a*), the remaining eight appear to be small-circle paths that have been distorted or flattened (Fig. 73*b*, *c*). Rotation paths of the sahlfold lineation in Riar basin were observed in four discfolds in amphibolite on Hemre Kam. They, too, appear to be distorted small-circle paths (Fig. 73*d*, *e*). The small-circle rotation paths centered about β's indicate folding by flexural slip. The distortion of the paths in the folds on Svahø suggests flattening subsequent to rotation. Although the strain involved in such

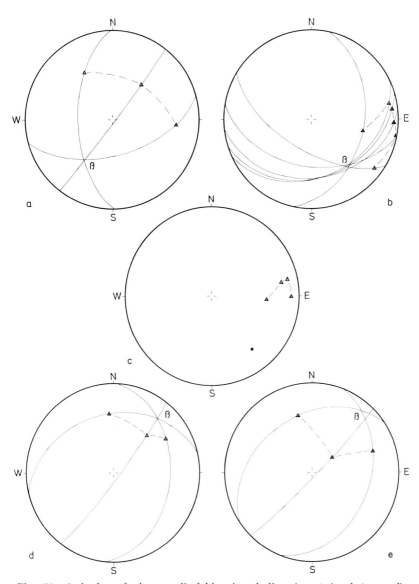

Fig. 73. Attitudes of the pre-discfold mineral lineation (triangles) on five discfolds. Solid great circles: foliation. Dashed great-circle arcs: approximate continua of lineation attitudes. *a, b,* and *c* Measured in flagstone on Svahø. *d* and *e* Measured in amphibolite on Hemre Kam

"flattening" is impossible to specify quantitatively from the data at hand, it is possible to conclude that shortening occurred across the short axes of the elliptical rotation paths, i. e. parallel to a steep, east-west trending axis, relative to elongation along the long axes of the paths, roughly parallel to the north-south, horizontal axis (Fig. 73 b, c). Since the axial planes of the discfolds on Svahø have shallow dips, the axis of shortening so deduced makes a large angle with the axial surfaces, and the axis of elongation approximately parallels them.

It is concluded from the two foregoing lines of evidence that disc-folds developed initially as flexural-slip folds with parallel geome-try and were flattened subsequently, perhaps perpendicular to their axial planes, as deformation proceeded. It is conceivable that the incip-ient schistosity that parallels the axial surfaces of some discfolds de-veloped during the later, flattening stage.

Slip-Line Orientations in the Svahø Border Zone

Discfold Movement

The planar distribution of discfold axes within the Svahø zone parallels the regional attitude of the foliation (Figs. 66b, 72b). It fol-lows, therefore, that discfolds developed in effectively planar folia-tion, within the limbs of earlier, isoclinal folds, as if no previous fold-ing of the layering had occurred there. In accordance with the con-clusion of the foregoing section, the discfolds are thus interpreted to be flexural-slip folds that developed in planar layering. As discussed in an earlier chapter (pp. 38—42), a separation angle defined by such folds confines the slip-line orientation under which the folds developed, assuming σ_2 parallel to layering.

Perhaps it should be emphasized that the slip-line orientations re-ferred to here with regard to flexural-slip folds are not the directions of slip between layers within the limbs of the folds. Rather, they are the orientations of the lines of relative slip between the sequence of layers that undergoes folding and the adjacent layers that do not fold.

Throughout the Svahø zone, groups of discfolds define separation angles oriented like the one in Fig. 72a. Slip-line orientations deduced from the separation angles trend uniformly east-west and plunge shallowly to the east, parallel to the regional foliation. A single sense of shear is compatible with the distributions of asymmetry of most of the groups of folds, shown in the composite diagram in Fig. 72b; the sense of shear indicates upward and westward movement of upper layers relative to lower ones, parallel to foliation. It is noteworthy that the same sense of shear is compatible with the discfolds observed in the greenschist rocks east of the Nonshø border fault, though the separation angle is large and the slip-line orientation is not well defined (Fig. 72c). The opposite sense of shear is compatible locally with folds on westernmost Svahø and Hammerhø, near their junction with Tyrikvamfjellet, and on southern Svahø and western Nonshø. These are the only slip-line data that pertain unequivocally to the movement in which the discfolds developed.

A case can be made for the proposition that the large flexures in the Svahø zone are simply the macroscopic equivalents of discfolds. Comparison shows that the descriptive properties of the flexures are the same, qualitatively, as the properties in Table 9. Because the flexures share the same relationships to other structures as the discfolds, they are probably of the same generation. For these reasons, they are thought also to be flexural-slip folds that developed in effectively planar layering, and therefore the separation angle they define (Fig. 69) can be interpreted like those of discfolds. The separation angle in Fig. 69 has the same orientation as those defined by discfolds in the Svahø zone (Fig. 72a, b), and the sense of shear compatible with the asymmetry patterns corresponds to the sense of shear compatible with most discfolds in the zone.

Other Movement

Three additional slip-line solutions, which may not relate directly to discfold development, were obtained in the Svahø zone:

On Svahø the hinge of a Lang tjern fold is rotated through 153° within the later fold shown in Plate 21. The rotation path closely approximates a plane, the pole to which makes a large angle with the later fold axis (Fig. 74a). The intersection of the plane of rotation with the axial plane of the later fold is a solution for the slip-line orientation by which the later fold developed, assuming folding by slip. Unfortunately, this fold has not been identified by type or generation.

Two solutions were obtained from the interference structures on Høghø. The hinge line of an early fold is rotated 159° in a somewhat planar path within a canoe-shaped antiform (group II, Fig. 13) resulting from superposition by a Høghø fold. Assuming folding by slip, the intersection of the approximated plane of rotation with the axial plane of the Høghø fold is taken as a slip-line solution (Fig. 74b). On the same assumption, the polygon described in spherical projection by attitudes of compositional layering about an antiform with convex profile (group II, Fig. 13) is taken as another solution (Fig. 74c).

Pl. 21. Superposed folds in marble and mica schist on Svahø (Fig. 74*a*). *Top*. View of the later fold at a low angle to its fold axis. Hinges of Lang tjern folds rotated during later folding can be seen at center, right. *Bottom*. View nearly at right angles to the fold axis of the later fold. Rotated hinges of Lang tjern folds are visable near the hammer pick, both straight (immediately to the right) and folded almost isoclinally (farther to the right)

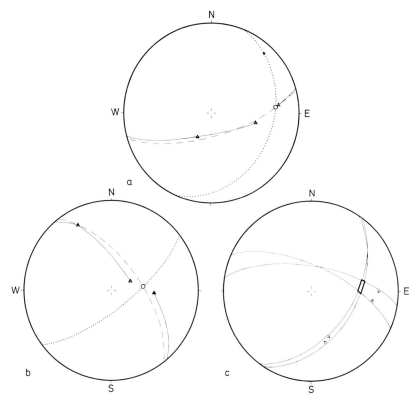

Fig. 74. Slip-line solutions obtained from folds other than discfolds within the Svahø border zone. *a* Axes (triangles) of a Lang tjern fold rotated during later folding on Svahø. Solid great-circle arcs: approximate continua of Lang tjern fold axes. Dashed great circle: plane approximating the Lang tjern fold axes. Dotted great circle and dot: axial plane and axis of the later fold. Open circle: slip-line solution for the later fold. *b* Axes (triangles) of an early fold rotated during Høghø folding, eastern Høghø. The structure is a canoe-shaped dome. Solid great-circle arcs: approximate continua of early fold axes. Dashed great circle: plane approximating the early fold axes. Dotted great circle: axial plane of interfering Høghø fold. *c* Attitudes of foliation (great circles) measured on an interference dome on eastern Høghø. Arrows indicate to which side of the foliation the stacking axis of the basin is oriented. They determine the polygon (heavy lines) that encloses the slip-line orientation (Høghø)

Summary of Relationships between Fabric and Kinematics

The slip-line solutions in Figs. 69, 72*b*, and 74 are summarized in Fig. 75. The separation angles from the composite diagrams of disc-folds and macroscopic flexures are nearly identical in orientation. The other three solutions are similar to the separation angles in strike, but different in plunge by as much as 30°; the differences in plunge reflect local variations in the dip of the homocline. From these data it is concluded that the relative slip recorded by discfolds within the Svahø zone is directed east-west, plunging about 30° to the east, and

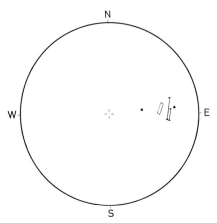

Fig. 75. Composite diagram of slip-line solutions from the Svahø border zone. Great-circle arcs: separation angles of discfolds (Fig. 72b) and macroscopic flexures (Fig. 69). Dots and polygon: solutions from interference structures not involving discfolds (Fig. 74a, b, and c)

that the directions of relative slip recorded locally by other folds in the zone is roughly the same. No widely divergent slip-line orientations were observed.

The Svahø zone, therefore, is characterized by an approximate coincidence of important fabric and kinematic axes of most of the generations of structures observed there. Linear preferred orientations of Lang tjern and Høghø fold axes (fabric b's), the linear orientation of the prominent mineral lineation (Fig. 67b), and the kinematic f axes of Høghø folds, discfolds, and macroscopic flexures (Fig. 75) share an east-west orientation. However, no principal fabric or kinematic axis of the Hammerhø synform and the system of mesoscopic folds that relates to it is known to share this orientation.

The histograms in Fig. 76 present the divergence between the slip-line orientation of discfolds and (a) attitudes of the prominent mineral lineation and (b) discfold axes. The histograms employ the center of the separation angle in Fig. 72b as the slip-line orientation and the lineation attitudes and discfold axes in Fig. 67b and 72b as the observations. Although the center of gravity of the point maximum of lineation attitudes in Fig. 67b is less than 10° from the slip-line orientation, the mode in histogram a falls in the 10°—15° class interval. As expected from Fig. 72*b*, the distribution of discfold axes relative to the slip-line orientation in histogram *b* displays a complete 90° spread without a significant mode. Differences in frequency in the histogram are not considered significant, however, because during the data-collecting an attempt was made at each individual locality to sample as wide a spread of axes as possible rather than to collect a sample representative of the natural distribution.

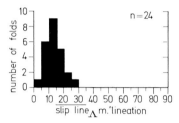

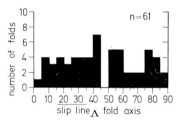

Fig. 76. Histograms showing the divergence between the orientation of discfold slip lines and *(a)* attitudes of the prominent mineral lineation, and *(b)* discfold axes

Nappe Emplacement

No direct evidence was recognized that bears on the sense of motion along the Nonshø border fault. In the absence thereof, the shear couple compatible with the asymmetry distributions of macroscopic flexures and of most groups of discfolds is used provisionally. The sense of shear recorded by these folds is westward and upward movement of the greenschist terrane of the Trondheim basin relative to the higher-grade rocks of Trollheimen. On the assumption that the folds formed under the same stress conditions as the fault, the shear recorded by the folds should be the same as the relative movement on the fault because the fault surface and the foliation (the surface of slip for the folds) have nearly the same orientation. Following this line of reasoning, the Nonshø fault is considered to be a thrust fault. The same reasoning applies for all possible faults within the Svahø zone; the folds indicate thrust movement.

The opposite sense of shear recorded by discfolds in a few localities within the Svahø zone appears for the most part to be insignifi-
cant to the regional geology and may simply represent the secondary
shear couple generated on the short limbs of larger folds. However,
the rare discfolds that are found on the lower, western slopes of Svahø
and Hammerhø, just east of the proposed southward extension of the
Grønhø fault, also display the opposite asymmetry distribution. In
the absence of other evidence, they are interpreted to indicate that the
sense of motion on the Grønhø fault was normal rather than thrust.

Nappe tectonics are suggested by: (1) the repetition of rocks of the
Gjevilvatn group structurally above rocks of the Blåhø group in the
eastern homocline, (2) the absence of a major hinge within the Blåhø
rocks on Tyrikvamfjellet and Høghø, (3) the structural distinctiveness
of the Svahø zone from the Trondheim basin and from the rest of
Trollheimen, (4) the deduced normal-fault character of the zone's
western boundary and thrust-fault character of its eastern boundary,
and (5) the presence of the Hammerhø synform. It is concluded, therefore, that the rocks in the Svahø zone compose portions of one or
more recumbent nappes that originated elsewhere and moved into
their present position.

The separation angles of discfolds and macroscopic flexures can be
taken as a manifestation of the shear couples set up during emplacement of the rocks in the Svahø zone. Therefore, the east-west slip-
line orientation represents the line of transport of the nappe(s) relative to adjacent rocks of the Trondheim basin and the rest of Trollheimen. The senses of motion deduced for the faults bounding the Svahø zone indicate that the direction of transport was eastward. Thus
the origin of the nappe(s) is fixed somewhere to the west, presumably
beyond Riar basin, which also contains the remnant of a nappe —
perhaps even the same one as in the Svahø zone, as reconstructed in
Fig. 48.

The following model, based for simplicity upon the emplacement
of a single nappe, is offered in explanation:

Let us suppose that somewhere to the west a large nappe developed, whether by upward flow and subsequent lateral spreading from
a "root zone" or simply by slumping of an uplifted and unstable geo-
synclinal pile. This process was recorded in the rocks by several generations and types of structures now seen as Lang tjern, Høghø, and
Hammerhø synform folds, pervasive foliation and mineral lineation,
and boudinage. As the nappe flowed eastward down a gentle incline
and plunged beneath the already recrystallized greenschist carapace
of the Trondheim basin, it thickened and its foliation became crumpled

due to headward resistance and the continual push of additional rock arriving from its source.

The block diagram in Fig. 77 is a generalized reconstruction of the Svahø zone according to this model. The front edge of the block is oriented east-west and passes from central Tyrikvamfjellet through Hammerhø to Nonshø. The diagram shows schematically the relations between the shear couples generated by the plunging nappe and

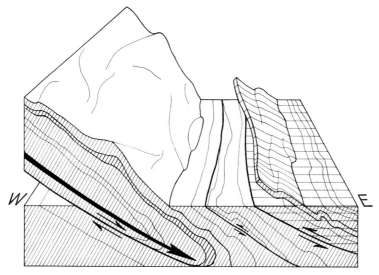

Fig. 77. Schematic block diagram showing the general relationship between a plunging nappe and resulting thrust faults and discfolds. The front face of the block passes from central Tyrikvamfjellet through Hammerhø to Nonshø. Horizontal ruling designates the greenschist terrane

the resulting faults, macroscopic flexures, and discfolds. As the nappe moved into its present position, its lower limb became increasingly attenuated and sheared, but the upper limb thickened in folding, accommodated by upward movement of the greenschist cover. Disc-folds and macroscopic flexures, their axes in all orientations within the foliation, were produced principally within the upper limb of the flowing nappe. Continued push of the more rootward portions of the nappe caused the forward portion to deform homogeneously so that early-formed folds became somewhat flattened and an incipient mineral lineation and schistosity developed locally.

The greenschist rocks were passive while the nappe beneath flowed into place. During an early phase of the deformation, they became separated from the more active, high-grade rocks along a shear zone, the Nonshø border fault, and they reflected the deformation below the surface by mild crumpling, the formation of discfolds whose axes are oriented nearly perpendicular to the stream lines, and by the local formation of a weak mineral elongation also perpendicular to the stream lines. The absence of a regional mineral lineation in the greenschist rocks is compatible with the conclusion that the greenschist terrane did not go through the same prolonged flow as the high-grade terrane, but instead floated passively above it. This is similar to the configuration seen by *Haller* in the Caledonide structures of East Greenland (1955a, pp. 42—44, Fig. 11; 1955b, p. 282, Fig. 2).

Flow Environment of the Discfold Facies

Like the sahlfold and norfold facies, the discfold facies is an assemblage of mesoscopic structures. It comprises discfolds, the poorly developed schistosity that parallels the axial surfaces of some discfolds, and the weak mineral lineation that parallels fold axes in the hinge areas of some discfolds.

Relative Orientations of Fabric and Kinematic Axes

Fold axes of discfolds describe the fundamental fabric plane, *ab*. Commonly, as in Fig. 72*b*, designation of the *a* and *b* axes within that plane is unwarranted from the distribution of fold axes. The designation would probably be pertinent in the presence of orientation data of axial surfaces, but such data were not collected during the present study. Where fold axes define a point maximum that does not spread into an entire plane, as in Fig. 72*c* and *d,* the center of gravity (β) of the maximum is defined as *b*, with the *a* axis 90° away. Fabric *c* is perpendicular to *ab*.

In a layered sequence undergoing local flexural-slip folding, slip occurs between the layers that flex and those that do not. This kind of relative movement of some layers past others can be described in the terminology of Chapter IV as velocity gradient flow. Where the folds develop in previously planar layers and their fold axes exhibit a planar distribution, as is the case with discfolds, the plane of axes (*ab*) parallels the plane of layering (foliation), which is also the principal plane of slip (cf. pp. 50—51). The principal plane of slip in velocity gradient flow is termed the principal stream surface, of which the flow axes are *df*. Thus kinematic *df* parallels fabric *ab* for the discfold facies. It follows also that their poles, kinematic *e* and fabric *c*, are parallel.

Where the asymmetry patterns of the folds define a separation angle and are therefore compatible with a single shear couple, which is also the case with discfolds, the line of action of the shear couple is confined by the separation angle, as already discussed in this chapter. The line of action of the shear couple which is the principal line of slip in velocity gradient flow, is termed the stream line, f. Kinematic f parallels neither fabric a nor b in all groups of discfolds (cf. Fig. 76b), although in some groups, such as those in Fig. 72c and d), kinematic f may parallel fabric a.

Flow Characteristics

Considerable east-west elongation of rocks in the Svahø zone is demonstrated by several strain indicators: elongate augen with long axes in the east-west direction, boudinage with boudin axes north-south, and Lang tjern and Høghø folds with long, straight hinge lines and strong linear preferred orientations of fold axes in the east-west direction. These structures are prediscfold, however, and as such record the strain of the rocks during the development and eastward flow of the nappe(s) rather than during the discfold event of emplacement. Physical characteristics of the discfolds themselves appear to be the only indicators of the strain accomplished during discfold time.

The production of flexural-slip folds with fold axes in all orientations within the plane of layering implies shortening in all directions within the layering. Both the curved hinge lines of discfolds and the absence of boudinage from the discfold facies are compatible with such shortening. Therefore, with reference to the principal flow axes of Fig. 25, shortening appears to have occurred parallel to both d and f. This necessitates elongation parallel to e, assuming incompressibility of the rocks. This conclusion is supported by the fact that the thickness of a layered body undergoing flexural-slip folding must increase to accommodate the folding. Thus the flow in which discfolds developed appears to have been compound, both converging with respect to d and diverging a greater amount with respect to e.

With the addition of the velocity gradient across e, discussed in the foregoing section, the essential characteristics of the flow seem to have been identified. The flow may have had other characteristics, such as minor rotation or a subordinate velocity gradient across d, but no evidence of them has been recognized. It is concluded, therefore, that the flow in which the discfold facies developed was a type of compound, velocity gradient flow, namely cdg. At present, we do not know enough about the homogeneous strain in which the discfolds were flattened to speculate meaningfully upon the characteristics of the later stage of flow.

IX. The Concept of Strain Facies

For our present purpose it is necessary to decide whether the three structural assemblages described from Trollheimen's rocks — the sahlfold, norfold, and discfold facies — are significantly different from each other physically. These assemblages are composed of cleavages, mineral lineations, elongate augen, folds, and boudins. Of these features, folds display the most obvious and varied physical differences and are thus the most useful for characterizing the assemblages. We begin, therefore, by examining the three fold types.

Comparison of Three Fold Types

Sahlfolds, norfolds, and discfolds are described in Tables 4, 6, and 9 as they appear in a single, uniform rock type. It is expected that this restriction has eliminated from the descriptions all effects that might result from the varied responses of different rock types. Thus any physical difference revealed in a comparison of the folds so described is more likely to reflect a difference in the modes or environments of the development of the folds.

Nevertheless, this restriction has not eliminated the effects that might result from a variable response of the single rock type itself in different local settings (i. e. different boundary conditions). For example, the strain seen as folds in a single rock type (1) in a thin layer, (2) at the edge of a thick layer, and (3) within an infinitely thick layer, all undergoing quantitatively similar flow, would be somewhat different because of the different degrees of influence of adjacent rock types upon the behavior of the rock type under consideration. A variation in adjacent rock types would also vary the behavior of the rock type concerned. Nevertheless, folds were studied in Trollheimen without regard to such local factors, in general, and their effects are expected to be incorporated into the tabulated descriptions. In the absence of pertinent data, however, such effects are ignored on the assumption that they are small and of equal importance to the descriptions of all three fold types.

Sahlfolds and Discfolds

Sahlfolds and discfolds are easily discernible as separate types. None of the twelve properties in Tables 4 and 9 are the same (pp. 70, 158). Obvious differences are numerous (e. g. similar vs. parallel geometry, long and straight vs. short and curved hinge lines, linear vs. planar distributions of fold axes), and even with regard to the properties whose ranges overlap, the mean or characteristic values are decisively different (e. g. cylindrical vs. irregular cylindricity, 1.2 vs. 0.5 mean H/W, 13.0 vs. 2.6 mean D/W).

It is possible to argue, however, that sahlfolds and discfolds developed as a single fold type but that one, the other, or both were later modified to their present forms. The conclusions reached earlier that sahlfolds may have been modified severely in extended convergent flow and that many discfolds were somewhat distorted (or "flattened") late in their development are compatible with such an argument. Perhaps the most logical hypothesis in this vein would be that sahlfolds developed from discfolds (or protodiscfolds) while undergoing the convergent flow recorded in Riar basin. During this transformation, the folds would have become flattened across their axial surfaces and greatly elongated within them, so that their profile geometry would begin to appear similar in type, their hinge lines would be stretched straight and their cylindricity become higher, their fold axes would approach a linear preferred orientation, their height-width and depth-width ratios would increase, and a pervasive schistosity and mineral lineation would develop.

Although on cursory examination this may seem an acceptable hypothesis, upon closer examination it is seen to fail. A number of technical discrepancies can be listed, such as the fact that the profile geometry of sahlfolds is strictly similar, not flattened parallel, and the strain associated with the flow in Riar basin was apparently not extreme enough to have produced nearly so concentrated a linear preferred orientation of fold axes from a truly planar distribution (pp. 112—113), but the telling discrepancy is found in the physical characteristics of certain individual folds. In virtually all groups of sahlfolds, individuals are found with nominal height-width ratios of 0.1 but with maximum heights of negative values and widths degenerate to the spacings between axial surfaces (Fig. 3b). In accordance with the hypothesis it is expected that sahlfolds with such low height-width ratios — well below the mean value for discfolds — would be the ones that had been modified least and would thus share other properties with discfolds. Nevertheless, they do not; they are similar folds with high cylindricity, straight hinge lines, and depth-width ratios above the highest values observed in discfolds. In short, they have all the characteristics of typical sahlfolds except height-width ratios near 1.0. These folds cannot be accounted for by the hypothesis. For these reasons, this and other hypotheses of a common origin for the folds are discounted.

Norfolds and Discfolds

Norfolds and discfolds share the same range and mean of height-width values (Tables, 6, 9; pp. 115, 158). Individuals of both types display weak schistosities parallel to axial surfaces and weak mineral lineations parallel to fold axes in hinge areas. The two fold types also share the same range of cylindricity, though a typical norfold exhibits higher cylindricity than a typical discfold. The range of variation in length and character of norfold hinge lines includes that of discfolds, though norfold hinge lines are somewhat longer and straighter on the average. Nevertheless, though similarities in the fold types do exist, they are outweighed by the differences. Norfolds are similar in profile geometry, but discfolds are parallel. Norfolds characteristically display sharp hinges and straight limbs, but discfolds display rounded hinges and limbs. Depth-width ratios of norfolds are high, but those of discfolds low. Fold axes of norfolds define linear distributions, but those of discfolds typically define planar distributions. Furthermore, norfolds in highly schistose rocks exhibit along their axial surfaces a well developed slip cleavage, which is absent in discfolds in all rock types.

The sharing of low height-width ratios argues against the possibility that norfolds developed from discfolds (or early, preflattening discfolds) by undergoing additional strain. Any incremental strain that might significantly increase the depth-width ratios and alter the profile geometry from parallel to apparently similar would also increase the height-width ratios. The opposite process of producing discfolds from norfolds by somehow changing geometric type from similar to parallel and *decreasing* the depth-width ratios while maintaining the same ratios of height to width seems impossible. Therefore, the probability of these two fold types having evolved from a single type is considered remote.

Sahlfolds and Norfolds

Differences exist between sahlfolds and norfolds, though they are not so numerous or obvious as those observed between either of these fold types and discfolds. Certain differences of degree, when considered together, however, constitute a significant gap between the two (Tables 4, 6; pp. 70, 115); broadly curved limbs compared to typically straight limbs, long and straight hinge lines compared to moderately curved hinge lines of intermediate length, high compared to moderate cylindricity, and high compared to low height-width ratios. In addition, where observed in highly schistose rocks, sahlfolds display a well developed schistosity parallel to their axial surfaces, but norfolds display an equally well developed slip cleavage.

Although the decision is not so clear as for the preceding pairs, these two fold types are also considered to have had separate paths of origin. If the sahlfolds had developed from norfolds — the easiest imagined of the possible paths — the axial-surface slip cleavage would have had to turn into schistosity as the norfolds became sahlfolds. Nevertheless, because the mechanics of slip cleavage and schistosity as understood today are almost totally different, it seems unlikely that one would be a stage in the development of the other. The existence in Trollheimen's rocks of various stages of development of schistosity and slip cleavage as individual structures, but the absence of any gradational stage or combination of parallel schistosity and slip cleavage, argue further against their being sequential products of a single path of development. Of course it could be suggested that sahlfolds developed only from norfolds without a slip cleavage, but such a suggestion appears *ad hoc*. Most groups of sahlfolds or norfolds in Trollheimen are composed of individuals in every stage of growth, as characterized by height-width ratios, but no individuals or groups were recognized as intermediates between sahlfolds and norfolds. Therefore, these two fold types are not likley candidates for sequential products of a growth series any more than their associated schistosity and slip cleavage are.

Rationale for Strain Facies

It is concluded from the foregoing comparison that sahlfolds, norfolds, and discfolds as viewed in Trollheimen today are significantly different physical types that did not develop from each other or from one or more common prototypes. Therefore, the three structural assemblages — the sahlfold, norfold, and discfold facies — each represented by a different fold type, are also significantly different from each other. The presence of boudinage and elongate augen in the sahlfold facies but their absence from the remaining two facies, and the presence of slip cleavage in the norfold facies but its absence from the remaining two, corroborate the conclusion that each of the assemblages is physically distinct.

The geometric flow types deduced from the properties and mutual orientations of the structures composing the assemblages and from their macroscopic structural environments are also fundamentally different from each other. The essential characteristics of the flow in which the sahlfold facies developed are convergence and a velocity gradient, c^2g. Those of the norfold facies are a velocity gradient and rotation, with convergence less important than for the sahlfold facies — c^2gr^2 or $pcgr^2$. The flow deduced for the discfold facies is compound, requiring both divergence and convergence in the direction of flow, as well as a velocity gradient — cdg.

The physical distinctiveness of each assemblage and the corresponding distinctiveness of the genesis of each assemblage, understood here in terms of simple flow environments, lead us to the concept of strain facies: *Physical properties of a structural assemblage express the conditions of its formation.* Practically anywhere a given assemblage occurs in Trollheimen, it looks the same and exhibits the same relative orientations of fabric and kinematic axes. This integrity of the assemblages further indicates that the facies concept is appropriate as a classification scheme for strain features. The concept implies, however, that all structural assemblages with the same appearance have the same genesis. The following examples are presented to show that this relationship appears to obtain in nature.

(1) The folds in a tundra landslide were considered at length in Chapter III to introduce the concept of the separation angle. Comparison of the physical properties of the tundra folds with those of discfolds in flagstone reveals a remarkable similarity (pp. 33—35; Table 9). Six of the twelve properties on the checklist are identical: parallel geometry, small to large radii of curvature in hinge areas and broadly curved limbs, short and curved hinge lines, cylindroidal and irregular cylindricity, planar distributions of fold axes, and normal asymmetry with small separation angles. Four of the properties — the ranges and means of height-width and depth-width ratios — are qualitatively the same; the extreme difference in the materials that display the folds — tundra sod and quartz schist — and their different effective viscosities during folding should account for the minor quantitative discrepancies in these values. Because the tundra sod displays no counterparts to the metamorphic cleavages and mineral lineations in the flagstone, the remaining two properties on the checklist cannot be compared.

Despite the fact that the tundra folds are flexures instead of flexural-slip folds and despite the extreme differences in materials and metamorphic grade, the tundra folds are classified in the discfold facies on the basis of the foregoing comparison. We should expect from the facies concept, therefore, that the geneses of these two sets of structures would have certain similarities. Such can be seen in their deduced flow environments, which are both cdg (cf. pp. 58, 173).

(2) In a discussion of the anticline hypothesis for the form of the flagstone "handfold" and Riar basin, folds in the Grand Saline and Jefferson Island salt domes were compared with sahlfolds (pp. 100–103). Ten descriptive properties on the checklist were concluded to be the same for the two sets of folds. Only one property is different; the salt folds are displayed by compositional layering and not by foliation. This discrepancy may indicate that the structural history of the salt prior to the folding under consideration was different from the presahlfold history of the flagstone, or it may simply reflect the individual behavior of different minerals undergoing metamorphism. The remaining property, cylindricity, cannot be compared because of insufficient knowledge of the salt folds. Despite extreme differences in rock type and conditions of metamorphism, the salt folds can be classified in the sahlfold facies on the basis of their close similarity in form to Trollheimen's sahlfolds.

Some of the characteristics of the flow in which the salt folds developed can be identified well enough for the purpose of comparison (*Balk*, 1949, 1953). It is clear that the salt within such upright domes as Grand Saline and Jefferson Island flowed upward, so that the stream lines, f, were nearly vertical, and that the linear preferred orientation of fold axes is roughly parallel to that orientation. In published drawings, thicknesses of folded layers measured parallel to axial surfaces are constant across some folds but small at the hinges of others; this is inconclusive but permissible evidence that the profile geometry of the folds is similar and that the folds developed by slip, which is assumed in this discussion. In the terminology of Chapter IV, slip folding is a type of velocity gradient flow in which the axial surfaces of the folds are parallel to the principal stream surfaces, df. The gradient acts across e by definition, though following the argument of finite lengths of hinge lines for slip folds (pp. 42–44, 110), a slight gradient may also have existed across d, within df. Folding is thought to have been initiated as the salt moved toward and into the domes (*Balk*, 1949, Fig. 23C, pp. 1813–1816); such flow must have been convergent with respect to both the d and e axes, though local exceptions are conceivable. The flow was rotational, moreover, as the salt passed gradually from lateral or centripetal flow toward the domes, to nearly vertical flow within the domes. Rotation was primarily about d because the principal stream surfaces were approximately tangential to the dome walls, but some subordinate rotation may have occurred about e as well.

In summary, the essential part of the flow environment in which the salt folds developed was apparently c^2gr, though the general environment may have been $c^2g^2r^2$. The environment deduced for the sahlfold facies in Chapter VI is essentially c^2g, though possibly as complex as $c^2g^2r^2$. It would appear, therefore, that the two flow environments are similar.

These examples indicate that strain features in widely different materials, structural settings, and metamorphic grades that nevertheless display certain similarities in form also have significant similarities in genesis. They indicate the applicability of the facies concept to strain features.

Some General Remarks

The idea of applying the facies concept to strain features is not new with this book (cf. *Harland*, 1956; *Dunbar* and *Rodgers*, 1957, p. 136; *Arthaud* and *Mattauer*, 1969). Moreover, the thought that structures look as they do because of the ways in which they developed will not seem startling, especially to field geologists who have used this commonsense approach to structural interpretation for many years. One might wonder, therefore, why so much time and effort have been expended in formalizing strain facies. The answer must be, simply, usefulness.

Strain facies as outlined in this work is a descriptive scheme for classifying folds and their associated structures on the basis of fold style, which denotes all the descriptive properties of individual folds within a group, as well as the properties of the whole group. Thus a set of structures is classified by the total appearance of its constituent fold set. This avoids some of the serious pitfalls common to field studies of folded rocks, such as treating each individual fold as a unique species because no two folds look exactly alike, grouping all folds together because "a fold is a fold" regardless of appearance or because they share one or two physical properties, or grouping all folds together that display, for example, the same axial orientation. Classification by fold style is a natural one because style relates directly to the processes by which the structures developed.

Strain facies can be useful in structural interpretation. A kinematic or dynamic history known for one structural complex can be understood for another insofar as the complexes display the same facies. From the viewpoint of a metamorphic structural analyst, it is enlightening, for example, to compare much of the flow by which Riar basin subsided in Trollheimen with the flow of salt into certain upright salt domes; the basis for comparison, of course, is their common sahlfold facies. In a similar way, it should be possible to understand better the structural history of a more intricate, a less well exposed, or simply a poorly known area by analogy to another area characterized by the same facies. The knowledge that wherever the sahlfold facies is recognized, for example, the flow direction for the deformation approximately parallels the sahlfold axes can be an important tool for the field geologist. A strain-facies map compiled for an area such as a quadrangle, for a major complex such as a nappe, or even for a whole mountain belt could be useful in delineating the location and intensity of development and the interference of the various facies present as well as in understanding part of the history of strain undergone by the rocks in their different parts. If the metamorphic core zones of alpine-type mountain chains, for example, were found to display a facies typical of constriction, favorably oriented (like c^2, longitudinally directed), one might be better able to conclude that mountain building of this type involves shortening of the earth's crust.

The three facies defined from Trollheimen are summarized in Table 10. Their distribution and intensity of development within the area covered by Plate 1 are shown on the map in Fig. 78. It should be pointed out that this is a partial strain-facies map because it does not show the numerous other sets of strain features, including folds, that occur in these rocks, mentioned in earlier chapters; they have not been studied in enough detail. From reconnaissance *(Hansen, Scott,* and *Stanley,* 1967) it would appear that the three facies defined are common in deformed belts, but this will be known only after considerable detailed field work. Certainly other facies also occur commonly in nature, just as other important, geometric types of flow occur in nature. Problems of classification might be expected where one facies grades into another, if that happens, where a facies is closely similar but distinct from another facies (subfacies?), or where a facies is modified through subsequent unrelated deformation.

It seems in retrospect that this book is an essay in solving a total structural problem, involving regional structural geology of the classical type as well as strain analysis and kinematics. Particular attention has been paid to deducing slip-line orientations, kinematic data without which this attempt could not have been made, but a great deal of work is still necessary before even this aspect of the problem is well understood. Nevertheless, it is hoped that the general concept of strain facies and some of the attendant methods and conclusions will prove useful in future structural studies.

Table 10. *Summary of three strain facies*

Sahlfold facies	Norfold facies	Discfold facies
Strain features[a]		
Sahlfolds, schistosity, mineral lineation, boudinage, augen elongation	Norfolds, incipient schistosity (slip cleavage in some rocks), incipient mineral lineation	Discfolds, incipient schistosity, incipient mineral lineation
Characteristics of index folds		
Similar, harmonic, almost isoclinal; long and straight hinge lines, high cylindricity, strong linear preferred orientation of fold axes; well-developed schistosity[b] parallel to axial surfaces, prominent mineral lineation[b] parallel to fold axes (cf. Table 4)	Similar, harmonic, open; fairly straight hinge lines, moderate cylindricity, planar limbs and sharp hinges; linear preferred orientation of fold axes; slip cleavage[b] along axial surfaces in schistose rocks (cf. Table 6)	Parallel, disharmonic, open; short and curved hinge lines, low cylindricity; planar preferred orientation of fold axes; small separation angles (cf. Table 9)
Identification of fabric axes		
b parallels the linear preferred orientation of sahlfold axes; ab parallels the planar preferred orientation of sahlfold axial surfaces; c is the pole to ab	b parallels the linear preferred orientation of norfold axes; ab parallels the mean orientation of norfold axial surfaces; c is the pole to ab	ab parallels the planar preferred orientation of discfold axes; b parallels the center of gravity of the most dense cluster of discfold axes in ab; c is the pole to ab
Additional fabric relationships		
Mineral lineation and augen elongation parallel b; schistosity parallels ab; linear preferred orientations of boudin axes parallel ab, at high angles to b	Mineral lineation parallels b; schistosity and slip cleavage parallel ab	Mineral lineation parallels discfold axes, and schistosity parallels discfold axial surfaces; neither is coaxial with a, b, or c, except fortuitously
Relative orientations of fabric and kinematic axes		
b nearly parallels f; a nearly parallels d; ab parallels df; c parallels e	b and a make moderate to low angles with f and d, respectively; ab parallels df^c; c parallels e^c	ab parallels df; c parallels e; b ranges from perpendicular to nearly parallel to f, as does a
Flow environment		
Convergent, velocity gradient flow (c^2g), with minor rotation (possibly $c^2g^2r^2$)	Rotational, slightly convergent, velocity gradient flow (c^2gr^2 or $pcgr^2$; possibly $c^2g^2r^2$ or pcg^2r^2	Compound, velocity gradient flow (cdg)

[a] Those listed are the ones recognized in the metamorphic rocks of Trollheimen. Although only the types of fold named (the "index folds") are included in a given facies, the same facies in different rocks may comprise fewer, the same, or additional strain features.

[b] Metamorphic grade permitting.

[c] For norfolds produced by the slip mechanism.

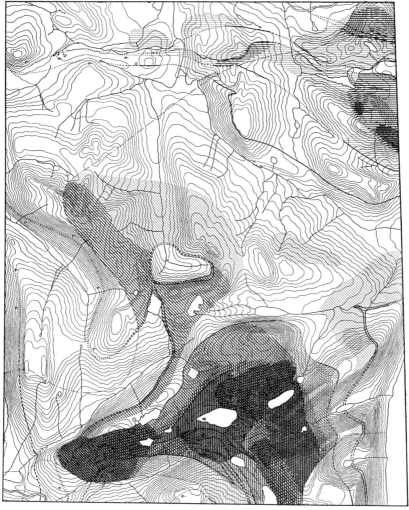

Fig. 78. Partial strain-facies map of southeastern Troll-heimen (cf. Pl. 1). The areas of development of the sahlfold facies are indicated by horizontal ruling, the norfold facies by oblique ruling, and the discfold facies by vertical ruling. An estimated intensity of facies development is shown by the relative weight of the ruled lines: light where poorly developed, medium where well developed, and heavy where very well developed and dominant. Penetrative structures of other generations (primarily earlier) and/or other strain facies are found throughout the map-area. Dashed lines indicate the limits of adequate observation.

References

Arthaud, F., and *M. Mattauer:* Présentation d'un nouveau mode de description tectonique: la notion de sous-faciès tectonique. C. R. Acad. Sci. Paris 268, 1019—1022 (1969).

Balk, R.: Structure of Grand Saline salt dome, Van Zandt County, Texas. Bull. Am. Assoc. Petrol. Geologists, *33*, 1791—1829 (1949).

— Salt structure of Jefferson Island salt dome, Iberia and Vermilion parishes, Louisiana. Bull. Am. Assoc. Petrol. Geologists, *37*, 2455—2474 (1953).

Barth, T. F. W.: Progressive metamorphism of sparagmite rocks of southern Norway. Norsk Geol. Tidsskr., *18*, 54—65 (1938).

Billings, M. P.: Structural geology. 2nd ed., 514 pp., New York: Prentice-Hall (1954).

Bloss, F. D.: Anisotropy of fracture in quartz. Am. J. Sci., *255*, 214—225 (1957).

Carey, S. W.: The Rheid concept in geotectonics. J. Geol. Soc. Australia, *1*, 67—117 (1954).

— Folding, J. Alberta Soc. Petrol. Geologists, *10*, 95—144 (1962).

Carstens, H.: Stratigraphy and volcanism of the Trondheimsfjord area. Norg. Geol. Undersøkelse No. 212*b*, 23 pp. (1960).

Carter, N. L., and *M. Friedman:* Dynamic analysis of deformed quartz and calcite from the Dry Creek Ridge anticline, Montana. Am. J. Sci., *263*, 747—785 (1965).

Clark, R. H.: A study of calcite twinning in the Strathavon marble, Banffshire. Geol. Mag., *91*, 121—128 (1954).

— , and *D. B. McIntyre:* The use of the terms pitch and plunge. Am. J. Sci., *249*, 591—599 (1951).

Dahlstrom, C. D. A.: Statistical analysis of cylindrical folds. Trans. Can. Inst. Mining Met., *57*, 140—145 (1954).

Davies, O. L.: Statistical methods in research and production. 396 pp. New York: Hafner (1957).

Dimroth, E.: Untersuchungen zum Mechanismus von Blastesis und Syntexis in Phylliten und Hornfelsen des Südwestlichen Fichtelgebirges. Tschermaks Mineral. Petrog. Mitt., 3rd Ser., *8*, 248—274 (1962a).

— Eine Theorie der Korngefügestatistik. Neues Jahrb. Mineral., Monatsh., 218—229 (1962b).

— Fortschritte der Gefügestatistik. Neues Jahrb. Mineral., Monatsh., 186—192 (1963).

Dobrin, M. B.: Some quantitative experiments on a fluid salt-dome model and their geological implications. Trans. Am. Geophys. Union, *22*, 528—542 (1941).

Donath, F. A., and *R. B. Parker:* Folds and folding. Bull. Geol. Soc. Am., *75*, 45—62 (1964).

Dunbar, C. O., and *J. Rodgers:* Principles of stratigraphy. 356 pp., New York: John Wiley & Sons (1957).

Elliott, D.: The quantitative mapping of directional minor structures, J. Geol., *73*, 865—880 (1965).

Flinn, D.: On folding during three-dimensional progressive deformation. Quart. J. Geol. Soc. (London), *118*, 385—428 (1962).

Gilmour, P., and *M. F. Carman:* Petrofabric analyses of Loch Tay limestone from Strachur, Argyll. Geol. Mag., *91,* 49—60 (1954).

Goguel, J.: Introduction à l'étude méchanique des déformations de l'écorce terrestre, 2nd ed. 530 pp., Paris: Imprimerie Nationale (1948).

Haller, J.: Der "Zentrale Metamorphe Komplex" von NE-Grönland, Pt. 1, Die geologische Karte von Suess Land, Gletscherland und Goodenoughs Land. Medd. Grønland, *73,* No. 3, 174 pp. (1955a).

— Die syn- und postorogenen Granite der ostgrönländischen Kaledoniden. Schweiz. Mineral. Petrog. Mitt., *35,* 280—286 (1955b).

Hansen, E.: Methods of deducing slip-line orientations from the geometry of folds. Carnegie Inst. Wash. Year Book *65,* 387—405 (1967a).

— Converging slip lines in Trollheimen, Norway. Carnegie Inst. Wash. Year Book *65,* 405—406 (1967b).

—, and *I. Y. Borg:* The dynamic significance of deformation lamellae in quartz of a calcite-cemented sandstone. Am. J. Sci., *260,* 321—336 (1962).

—, *S. C. Porter, B. A. Hall,* and *A. Hills:* Décollement structures in glacial-lake sediments. Bull. Geol. Soc. Am., *72,* 1415—1418 (1961).

—, and *W. H. Scott:* Real versus apparent displacement in slip folds. Carnegie Inst. Wash. Year Book *66,* 538—539 (1968).

—, and *W. H. Scott:* On the "drag folds" of *Van Hise* and *Leith* (1911). Carnegie Inst. Wash. Year Book *67,* 258—263 (1969).

—, *W. H. Scott,* and *R. S. Stanley:* Reconnaissance of slip-line orientations in parts of three mountain chains. Carnegie Inst. Wash. Year Book *65,* 406—410 (1967).

Harland, W. B.: Tectonic facies, orientation, sequence, style and date. Geol. Mag., *93,* 111—120 (1956).

Holtedahl, H.: Geological and petrological investigations in the north-western part of the Opdal quadrangle. Årbok Univ. Bergen, Naturv. Reeke, *1949,* No. 7, 60 pp. (1950).

Holtedahl, O.: Geological observations in the Opdal-Sunndal-Trollheimen district. Norsk Geol. Tidsskr., *18,* 29—53 (1938).

Howard, K. A.: Flow direction in triclinic folded rocks. Am. J. Sci., *226,* 758—765 (1969).

Knopf, E. B., and *E. Ingerson:* Structural petrology. Geol. Soc. Am. Mem. *6,* 270 pp. (1938).

Mackin, J. H.: Some structural features of the intrusions in the Iron Springs district. Utah Geol. Surv. Guidebook to the Geology of Utah, No. 2, 62 pp. (1947).

Martin, H.: The hypothesis of continental drift in the light of recent advances of geological knowledge in Brazil and in south west Africa. Trans. Proc. Geol. Soc. S. Africa, Annexure to *64,* 47 pp. (1961).

Matthews, D. H.: Dimensions of asymmetrical folds. Geol. Mag., *95,* 511—513 (1958).

McIntyre, D. B.: Note on two lineated tectonites from Strathavon, Banffshire. Geol. Mag., *87,* 331—336 (1950).

—, and *F. J. Turner:* Petrofabric analysis of marbles from mid-Strathspey and Strathavon. Geol. Mag., *90,* 225—240 (1953).

Muehlberger, W. R.: Internal structure of the Grand Saline salt dome, Van Zandt County, Texas. Bur. Econ. Geol., Univ. Texas, Rept. Invest., No. 38, 23 pp. (1959).

Muret, G.: Partie S. E. de la culmination du Romsdal, chaine caledonienne, Norvège. Intern. Geol. Congr., 21st, Copenhagen, 1960, Rept. Session Norden, Pt. 19, 28—32 (1960).

Nettleton, L. L.: Fluid mechanics of salt domes. Bull. Am. Assoc. Petrol. Geologists, *18*, 1175—1204 (1934).

O'Driscoll, E. S.: Experimental patterns of superposed similar folding, J. Alberta Soc. Petrol. Geologists, *10*, 145—167 (1962).

— Interference patterns from inclined shear fold systems. Bull. Can. Petrol. Geol., *12*, 279—310 (1964).

Page, B. M.: Gravity tectonics near Passo della Cisa, northern Apennines, Italy. Bull. Geol. Soc. Am., *74*, 655—671 (1963).

Parker, T. J., and *A. N. McDowell:* Model studies of salt-dome tectonics. Bull. Am. Assoc. Petrol. Geologists, *39*, 2384—2470 (1955).

Ramsay, J. G.: The deformation of early linear structures in areas of repeated folding. J. Geol., *68*, 75—93 (1960).

— The geometry and mechanics of formation of "similar" type folds. J. Geol., *70*, 309—327 (1962a).

— Interference patterns produced by the superposition of folds of similar type. J. Geol., *70*, 466—481 (1962b).

— Folding and fracturing of rocks. 568 pp. New York: McGraw-Hill Book Co. (1967).

Reynolds, D. L., and *A. Holmes:* The superposition of Caledonoid folds on an older fold-system in the Dalradians of Malin Head, Co. Donegal. Geol. Mag., *91*, 417—444 (1954).

Rosenqvist, I. T.: The Lønset anticline in the Opdal area. Norsk Geol. Tidsskr., *21*, 25—48 (1941).

Rouse, H.: Elementary mechanics of fluids. 376 pp. New York: John Wiley & Sons (1946).

Sander, B.: Zur Petrographisch-Tektonischen Analyse III. Jahrb. Geol. Bundesanstalt (Austria), *76*, 323—406 (1926).

Scott, W. H.: Experiments in flow deformation. Carnegie Inst. Wash. Year Book *67*, 251—254 (1969).

— , and *E. Hansen:* Movement directions and the axial-plane fabrics of flexural folds. Carnegie Inst. Wash. Year Book *67*, 254—258 (1969).

— , *E. Hansen,* and *R. J. Twiss:* Stress analysis of quartz deformation lamellae in a minor fold. Am. J. Sci., *263*, 729—746 (1965).

Skehan, J. W.: The Green Mountain anticlinorium in the vicinity of Wilmington and Woodford, Vermont. Vermont Geol. Surv. Bull., No. 17, 159 pp. (1961).

Stanley, R. S.: The bedrock geology of the Collinsville quadrangle. Conn. State Geol. Nat. Hist. Surv. Quad. Rept., No. 16, 99 pp. (1964).

— Bedrock geology of the southern portion of the Hinesburg synclinorium. *In* Barnett, S. G., editor, Guidebook to field excursions at the 40th annual meeting of the New York State Geological Association, May 1969, 37—63 (1969).

Strand, T.: The pre-Devonian rocks and structures in the region of Caledonian deformation. Norg. Geol. Undersøkelse, No. 208, 170—284 (1960).

— The Scandinavian Caledonides — A review. Am. J. Sci., *259*, 161—172 (1961).

— , and *P. Holmsen:* Stratigraphy, petrology and Caledonian nappe tectonics of central southern Norway; Caledonized basal gneisses in a northwestern area. Norg. Geol. Undersøkelse, No. 212 l, 31 pp. (1960).

Trusheim, F.: Über Halokinese und ihre Bedeutung für die Strukturelle Entwicklung Norddeutschlands. Z. Deut. Geol. Ges., *109*, 111—158 (1957).

Turner, F. J., and *L. E. Weiss:* Structural analysis of metamorphic tectonites. 545 pp. New York: McGraw-Hill Book Co. (1963).

Tuttle, O. F., and *N. L. Bowen:* Origin of granite in the light of experimental studies in the system $NaAlSi_3O_8$-$KAlSi_3O_8$-SiO_2-H_2O. Geol. Soc. Am. Mem. 74, 153 pp. (1958).

Van Hise, C. R., and *C. K. Leith:* The geology of the Lake Superior region. U. S. Geol. Surv. Monograph 52, 641 pp. (1911).

Watson, G. S., and *E. Irving:* Statistical methods in rock magnetism. Monthly Notices Roy. Astron. Soc., Geophys. Suppl., 7, 289—300 (1957).

Wegmann, E.: Beispiele Tektonischer Analysen des Grundgebirges in Finnland. Bull. Comm. Geol. Finland, 87, 98—127 (1929)

Weiss, L. E.: Fabric analysis of a triclinic tectonite and its bearing on the geometry of flow in rocks. Am. J. Sci., 253, 225—236 (1955).

— Geometry of superposed folding. Bull. Geol. Soc. Am., 70, 91—106 (1959a).

— Structural analysis of the Basement system at Turoka, Kenya. Overseas Geol. Mineral Resources (Gt Brit.), 7, 3—35, 123—153 (1959b).

— , and *D. B. McIntyre:* Structural geometry of Dalradian rocks at Loch Leven, Scottish Highlands. J. Geol., 65, 575—602 (1957).

White, W. S., and *R. H. Jahns:* Structure of central and east-central Vermont. J. Geol., 58, 179—220 (1950).

Author and Subject Index

Errata

Page 110, line 29: for *in* read *is*

Page 112, line 21: for *25°* read *35°*

Page 112, line 34: for *are* read *is*

Page 129, line 3: for *fingers* read *'fingers'*

Page 129, Fig. 57: disregard dotted line in upper histogram;
disregard entire lower histogram

Page 132, line 17: for *of norfold-facies structures* read *of the norfold facies*

Page 133, lines 19—20: for *sahlfold-facies structures* read *the sahlfold facies*

Page 133, line 38: for *sahlfold-facies structures* read *the sahlfold facies*

Page 133, line 42: for *structures of the norfold facies* read *norfold facies*

Page 157, line 32: for *Thickness* read *Thicknesses*

Page 165, line 40: for *group II* read *group I*

Page 167, caption to Fig. 74, line 12: for *basin* read *dome*

Minerals, Rocks and Inorganic Materials